OXFORD MEDICAL PUBLICATIONS

....

Caring for a Dying Relative

Caring for a Dying Relative

A Guide for Families

DEREK DOYLE

Medical Director
St Columba's Hospice
Edinburgh

Oxford New York Tokyo
OXFORD UNIVERSITY PRESS
1994

Oxford University Press, Walton Street, Oxford OX2 6DP
Oxford   New York   Toronto
Delhi Bombay Calcutta Madras Karachi
Kuala Lumpur Singapore Hong Kong Tokyo
Nairobi Dar es Salaam Cape Town
Melbourne Auckland Madrid
and associated companies in
Berlin Ibadan

Oxford is a trade mark of Oxford University Press

Published in the United States
by Oxford University Press Inc., New York

A catalogue record for this book is available from the British Library

Library of Congress Cataloging in Publication Data
Doyle, Derek.
Caring for a dying relative: a guide for families/Derek
Doyle.
Includes bibliographical references.
1. Terminal care. 2. Terminally ill—Home care. 3. Terminally
ill—Family relationships. I. Title.
R726.8.D679 1994   362.1′75—dc20   94-1328
ISBN 0 19 262487 3 (Pbk)
ISBN 0 19 262488 1 (Hbk)

Typeset by Advance Typesetting Ltd, Oxfordshire
Printed in Great Britain by
Biddles Ltd
Guildford & King's Lynn

Preface

Some years ago I was invited to deliver a lecture and given the title 'Caring for the dying: the ultimate challenge'. It was intended for doctors and nurses but could equally well have been prepared for lay people because they are the ones who always have been, and always will be, the principal carers. Most patients spend many weeks in hospital undergoing sophisticated tests and treatment but, contrary to what most people believe, most of the final year of life is spent at home under the devoted care of family and friends. This book is written for them—the untrained, untutored carers who so willingly shoulder immense responsibilities, doing something they have never done before, every action, every day coloured by the knowledge that soon they will lose the one they love.

As far as possible I use simple, everyday language instead of the mystifying jargon of the medical profession. The advice given is, where possible, simple and basic, highlighting what everyone can do in partnership with the professionals. It applies whether the patient is male or female, even where I have used male gender in the text.

I write from long experience in this work, both in hospitals and hospices, and in the community, enriched beyond words by the teaching and example of colleagues, nurses, patients, and their family carers. To them all, and in particular to my friends in the hospice movement and my secretary, Mrs Irene Turnbull, I express my profound debt.

The book is dedicated to all who care for the dying, who accept this challenge—the ultimate challenge.

Edinburgh   D.D.
April 1994

Contents

1

Introduction

'I am very sorry to have to tell you that your husband has cancer.'

The effect of these words can scarcely be described. Countless questions and thoughts flash through your mind. Can it be treated? Can it be cured? How long will I have him? Will he suffer? Where will he be cared for? Will I be able to keep him at home and, if so, how on earth will I cope?

The thoughts and unspoken questions are almost the same whether the patient is a husband or wife, a partner, a child of any age, an aged parent, or a dear friend: how will I cope?

They are the same questions whether the doctor speaks of cancer or any of the other life-threatening illnesses. The same self-doubt raises its head. Will I be able to care for him?

Perhaps months or years pass with a series of operations and treatments, some designed to prolong life, others to keep him as free as possible from pain and suffering. Eventually there comes the time long expected and feared beyond description—'I am afraid there is little more we can do. He is going to die and we must talk about what we can all do for him now'.

How am I going to care for him? How will I cope? What do I say to him? What do they expect of me? I feel so frightened, so unskilled, so utterly helpless.

This book is written for those who are asking these questions— the ordinary men and women whose relative or friend is dying and who ask 'How do I care for him?'.

It is a book *exclusively* for the so-called layman—not trained in medicine or surgery; for the person who may be capable and confident in their daily work but who now feels utterly helpless yet determined to care and cope in a crisis they have never before been through, surrounded by professionals who seem to speak a different language, and live in a world which is safe for them but apparently hostile for the carers.

No book can take away the pain or bring back happiness. All that is possible is to strengthen determination or to make some sense in a world which suddenly seems to have fallen apart. This book sets out to show how, in so very different ways, the untrained carer *can* cope, *can* care, and *can* do so more capably and lovingly than they had ever dared to think.

2

The evening of life

Most doctors at some time or other are called upon to care for the dying but there is now a branch of medicine devoted to it—palliative medicine. Like all medical specialties it has a defined remit but part of it has a particular relevance to this book and to all who turn to it for guidance and encouragement. It states, among other things, that specialists in palliative medicine should have as the focus of their care *the quality of life*. Many might have expected it to speak of the quality of dying, or the easing of unnecessary suffering, but instead it places the emphasis on quality of living. The distinction is an important one for us all.

When someone we care for is known to be coming to the end of life, it is natural that we should want to ensure that he or she will not suffer pain; that he will not have reason to fear choking, suffocating breathing, or any other terrifying feature we have come to associate with death; that he will not have to endure fear or loneliness. We look to our doctors and nurses to see to these things for they are the experts wherever they work. We put our relatives in their hands, disappointed that they cannot cure, but always hopeful that death itself will be peaceful, for it is a commonly held belief that it is death which Man fears more than anything else in life.

In fact Hippocrates, writing long ago, noted that 'Young men fear death, old men fear dying'. This still holds true today. For the young, death is an enemy which comes too soon, robbing them of their dreams, destroying their hopes of seeing their children growing up around them, and depriving them of any chance to make this world a better place for those children. For the older man or woman, death is not the enemy they fear—it is the process of dying, the suffering they may have to endure, and the dignity and the sense of usefulness which has given life meaning for them that they may so easily lose. Dying can mean frailty usurping strength and vitality, physical disfigurement and disability replacing athletic prowess and sexual attractiveness, company and social intercourse giving way to painful dependency on others, and often desolate loneliness.

Ask anyone approaching death, as I have done thousands of times, and they will rarely speak of death itself or any fear of it. They will tell of tiredness and weakness they had never thought possible; of how they hate being so dependent on others for their every need; of how much they still want and crave to feel needed in a world which seems so ready to forget them and continue without them. They will speak of thoughts and questions they have never had before—thoughts about life and its meaning, questions about God, the value of life now behind them, and whatever lies ahead.

All who care for the dying, whether they be the professionals or the family and friends, would do well to take note. Their task might seem to be simple nursing and attention to basic physical and psychological needs and, up to a point, they are correct. This book is designed to help them do just that, but their responsibility goes much, much further. Our task is to address a person's deepest needs, not merely those of the tired, frail, aching body.

One of the booklets recommended in the book list at the end of this book is entitled, 'Dying—the greatest adventure of my life', written just before he died by a young doctor with a deep Christian faith. The title is startling—'. . . the greatest adventure of my life'. He was not unique. He was not even unusual. Many thousands of people looking back on their recent experience only days before death have spoken of it in similar vein. Friendships have become richer and love even deeper; differences have been resolved, old feuds forgotten, and enemies forgiven; faith has grown and meaning been found for many of life's mysteries. For some, life only seemed to take on a meaning and have a purpose as it came to its end; others felt loved and needed for the first time. What a paradox! Life reaching its climax, not in social or intellectual achievement and success, but in death!

It would be easy to sneer cynically and say that such experiences are rare, but they are not. They are common. It would be natural for those who watch and wait to say that that is all very well for the patient, but what about the person who can only see a loved one leave them, helpless and lonely with little or nothing to live for: but they too are wrong. Just as the dying can 'grow' even as they die, so can those who tend them so lovingly. They, too, can learn and grow and, in years to come, look back on their time with satisfaction in what they managed to do and wonder at what they learned. Perhaps they saw love in a new light or realized how petty were most of life's squabbles; perhaps they learned about understanding and reconciliation, of sacrificial caring and of the immense strengths we all have within us, so often untapped and untested.

We live in a world which fears and tries to deny disability, in a world which puts a premium on health and strength and sees no possible good coming out of death and disaster. This is to forget our history. What of the musical genius of the deaf Beethoven or the tormented Tchaikovsky; what of the inspiration of Keller, born deaf, blind, and mute, or the conciliatory works of the holocaust survivors? Have we forgotten the Douglas Baders and the Leonard Cheshires of this world, the nameless, maimed heroes of our wars who came back to teach us how to care and make the world a better place? Everything they did and created stemmed from suffering and pain.

When we accept this challenge of caring for the dying, we are not merely doing our duty. We are committing ourselves to more than easing pain and suffering, more than sitting sadly by the bedside. We are dedicating ourselves to love in action—ready to receive as much as to give, prepared to learn and even change and be changed, ready to ponder anew what we mean by dignity and honesty, and ready to ensure that the one we love will, in some mysterious way, *live* until he or she dies, leaving us and the world better than when they entered it—hence this book.

3

Home, hospital, or hospice?

It is often said that, given the choice and all other things being equal, most people would like to die at home, cared for by loved ones, surrounded by their families and modest possessions each holding a memory. It's easy to understand why they say this. What is not so easy to explain is why fewer and fewer people are actually dying at home. Within living memory, 70 per cent of people in Britain died at home. Today the figure, certainly in most towns and cities, is around 25 per cent and steadily falling.

There is no simple explanation why this is happening because the reasons are many and complex.

Hospital

Understandably, hospitals have come to be equated with an abundance of skilled medical and nursing care, ready access to investigations such as X-rays, ultrasound scans, and computerized tomography, not to mention operating theatres, intensive care and life-support facilities, and laboratories. It is there that most serious illnesses are diagnosed, or at least confirmed; there where medical specialists work and where most patients with serious illnesses have spent weeks or months under their skilled care and then returned to a full life again. It's only natural that people generally rate hospitals highly and usually feel safe in them.

Home

Home may be familiar and usually infinitely more comfortable but it doesn't always feel as safe. It takes longer for a doctor to come when needed than it would in hospital. Some people believe, quite wrongly, that the family doctor has less knowledge and expertise than a hospital doctor when it comes to looking after the dying; in fact, the opposite is often the case. Community nurses may only be able to visit a few times a week and even when they come once or twice a day this visit must be short, as there are so many other calls upon their time. Relatives soon look tired and strained, especially when the illness is a protracted one and the end will mean death for the patient and grief for them. Sometimes there are not sufficient carers to provide around-the-clock care. Some are old or frail, some young and going out to work. Modern homes are rarely equipped with a room easily converted into a sickroom.

There are, however, more subtle reasons why home care so often seems an unsustainable ideal. We live in an age when people expect 'expert' or 'specialist' care; when relatives feel incapable of doing what is required because they do not feel they are qualified or trained, wrongly assuming that the care of the dying always needs special skills and equipment. This is simply not so, but many believe it. They are daunted by the responsibility of nursing or feeding the one they love, terrified that 'something will happen' by which they usually mean incontinence, bleeding, choking, fits, or death itself.

It sometimes seems to me that society itself has delegated its concern and responsibility for the dying to the experts—the doctors and nurses of our hospitals and hospices, as if people feel they should not be expected to take on this labour of love. One hears of neighbours and friends expressing shocked surprise that families are even contemplating home care, no matter how willing and capable they are, when there are hospitals and hospices provided for them.

Hospices

What are hospices? Hospices, in the modern sense of the word, are relatively new features in our society. Most have been developed only in the last decade of the twentieth century but are now to be found throughout the country. They are somewhat unusual in being a British concept now eagerly adopted and developed worldwide, with thousands to be found in North America, throughout the former Commonwealth, across Europe, and as far afield as Japan, Korea, Israel, Iceland, and the Commonwealth of Independent States, as the USSR is now called.

The word 'hospice' is now a hallowed one, accepted into our language, but to doctors and nurses a more accurate description (and the one usually used by them) is a palliative care unit. This better defines their role—the care of people with 'active, progressive, far-advanced illness who have a relatively short life left to them and their need to be guaranteed the best possible quality of remaining life', something achieved by highly skilled attention to every aspect of their physical, emotional, social, and spiritual suffering by a closely integrated team of professionals. They are there to relieve suffering, not to prolong life. They recognize they cannot cure but they can, in a thousand ways, continue to care. They set out neither to extend nor to abbreviate life.

Most hospices in Britain have in-patient beds, aim to rehabilitate home as many patients as possible if only for a few weeks or months, and are increasingly staffed by doctors specially qualified in this difficult and demanding work and nurses equally highly trained and committed to it. Not only are they small in-patient units but most have a home care service offering specialized support and advice for patients and families at home, and often there is a day hospice and facilities for out-patient consultations, with members of staff (or integrated teams) who advise in local hospitals.

There are still people who think of hospices as places where people go to die—'No one ever comes out alive'. They imagine them as depressing and dismal, the atmosphere one of doom and gloom, unmitigated sadness, and misery. *Nothing* could be further from the truth. Every year thousands of people go in to have specialist attention to relieve their suffering, whether it is pain, breathlessness, sickness, fear, or whatever, and then return home. Their underlying condition has not been cured; in fact it may not have altered in any way. In many cases the disease has probably continued to get worse while they were in the hospice because there was nothing to control it, but its effects have been 'palliated', hence the name of the specialty—'palliative medicine' or 'palliative care'. Sooner or later most people will have to return but many go in and out many times in that last year or so of life. Describing the atmosphere is not easy, particularly for someone like myself who has worked in hospices for so many years. Perhaps it's best described by some of the words most used by patients and their relatives—'safe', 'relaxed', 'friendly', 'homely', 'caring', 'nothing ever seemed a trouble' . . .

Lest this reads like an advertisement, it is important to stress that hospice care is not for everyone. Some people prefer to return to the hospital unit where they have become known over the years and be looked after by the nurses and doctors they have justifiably come to trust. Others feel that entering a hospice is too final, too open an acknowledgement that there is no cure. A few fear that going into a hospice will cut them off from other specialists, but happily of course this is not so. Specialist palliative care workers cooperate very closely with such fellow specialists as surgeons, oncologists, radiologists, and many others, with the result that hospice patients are always moving backwards and forwards between the hospice and radiotherapy and other services. Many patients are visited in the hospice by the oncologists and others who have known them for so long. Others are seen there for the first time by these specialists, called in by the hospice doctors to advise. Hospice care is not 'end care' but part of a continuum of care. Still, quite rightly, not everyone wants it nor should they be expected to commit themselves to another group of carers no matter how high their reputation, unless they want to.

Making the decision

By now it should be clear that 'home, hospital, and hospice' are not three unrelated options. It is not a matter of picking the best one but rather of making use of each for different needs at different times. Each offers something the others cannot offer as well as many of the things they have in common.

For sophisticated investigations and highly technical treatment at any stage in an illness, the place to be is in a hospital. When everyone acknowledges that cure is not feasible and the need is for relief of any kind of suffering, the hospice or palliative care unit is the place. At all other times, no matter how difficult or daunting the prospect may appear, home should be the preferred place.

Making the decision is not as difficult as it might appear. Guidance and recommendations will be offered by the general practitioner (advised if needs be by specialist colleagues) and a decision made with the help of family and friends. If the diagnosis or extent of the disease is in doubt, if an operation or other special procedure is needed—you go into hospital and after that might either return home directly or via a hospice. If the pain is proving difficult to control, the frailty is such that many pairs of hands are needed to help, or the carers need a respite break, then a stay in the hospice is a possibility.

Care at home is possible on very many more occasions than people realize. It does not depend upon whether or not the house is palatial or luxurious, nor on having skilled assistants on hand, nor on an abundance of special aids and equipment. It depends on there being enough carers willing to help, willing to share with and support each other, and a preparedness to use, listen to, and learn from, the professional helpers who are available.

Other chapters in this book will describe the many aspects of caring at home, the layout of a sickroom, and the ways of caring rather than merely coping. It only remains to repeat yet again—most people *do* want to be at home. We should all try harder to make it possible.

Perhaps the final words on this subject should be those so frequently said by families who did manage to care for a loved one at home until death: 'Yes, it was difficult and exhausting but it was a wonderful experience we shall never regret and, looking back, one of the most rewarding experiences of life.'

4

Working with the professionals

One of the themes of this book for carers is, 'You can do it!'. The medically unqualified person who wants to care for someone they love, wherever he or she is, but feels inadequate for the task, can do much more than they ever dreamt was possible. They can do so even more capably and adequately if they work with, and learn from, the professionals with whom they come in contact: doctors, nurses, therapists, social workers, and chaplains.

The first requirement is a commitment to care and to share in doing so with others, including the 'experts'. The second, I suggest, is the use of a small notebook. Use it every day, have it with you when you see the doctor, visit the hospital, or sit with the patient. Use the left-hand page for questions you must remember to ask and the opposite page for answers, advice, and instructions. At the back you can jot down the things you must remember to do but which can so easily be forgotten.

Such advice may sound trivial until you see the doctor and can't remember what it was you said you must ask him, and afterwards what advice he gave. People become forgetful when they're under stress and they come to think they are developing dementia. No doctor or nurse will laugh at a notebook being brought out, though they may look alarmed if told there is a pageful of questions to be answered in the next few minutes.

Remember the rule when caring for someone who is dying— 'Nothing is trivial'. If it puzzles you or troubles you, then ask! If you write it down, you may not need to ask that question again.

The different professionals

It is helpful to know the different roles of the professionals with whom you may come into contact. They are all different but complementary.

Doctors

The general practitioner (GP) or family doctor (surely a happier name) is responsible for all the medical care of people at home —health education, disease prevention, early diagnosis, basic investigations, choice of and referral to hospital specialists, care after discharge, inviting specialists to advise in people's homes, and terminal care. Hospital doctors are either specialists or (except for the most junior) training to be specialists. The senior ones are called consultants, while their trainees are registrars and senior registrars. The most junior doctors, usually resident in the hospital, are the house officers often referred to as 'housemen' or 'resident'. It has been said that GPs must know a little about a lot; and consultants a lot about a little! Their domain is the hospital with its sophisticated equipment and diagnostic facilities. When a patient is in hospital, he is the responsibility of the consultant (though often it is the registrars he will see most frequently) and when he goes home he is under the GP; but all concerned will endeavour to keep each other well informed and up to date.

Nurses

The senior nurse in a hospital ward is a 'sister' (or charge nurse if a man), the other registered nurses being called staff nurses. In addition, there are student nurses and nursing auxiliaries.

Nurses working with patients at home are called community nursing sisters. They are still often referred to as district nurses or health visitors, the former providing the 'hands-on' care of bathing, injections, dressings, enemas, and so on, the latter focusing on health prevention, inoculation, care of mothers and babies, and the elderly.

Another group to be found almost exclusively in Britain are the Macmillan nurses, so called because in the first few years of any new service employing them, their funding comes from the Cancer Relief Macmillan Fund (CRMF). They are usually very experienced community nurses who have undertaken further training in palliative care nursing. Others are counsellors, paediatric specialists, or liaison nurses between hospitals and the community. Other cancer nurses include Marie Curie nurses employed by local authorities with funds from Marie Curie Cancer Care. They provide an invaluable service, sitting in with patients at home principally at night but in some cities a day service is also available. The frequency of visits by nurses is decided by them, not the doctors, but in caring for the dying they all work very closely, deciding among themselves what input is needed and what outside help should be called in to the home.

Occupational therapists

Employed by social work departments are occupational therapists working in patients' homes and in hospitals themselves. Their job is to help people, whatever their illness, to live life to the full no matter how disabled they feel and limited their abilities. They assess for and provide aids and equipment for the home, ranging from elevated toilet seats, handrails, and communication systems, to modified cutlery, crockery, and kettles—all to assist in what they describe as 'activities of daily living' (ADL). Their special skills are called on by doctors and nurses wherever needed.

Physiotherapists

In both hospital and community, there are physiotherapists, trained to rehabilitate—that is to help patients to lead the fullest and most useful life in spite of illness and disability. They can teach people how to sit properly, stand, move and breathe correctly, getting them back on their feet, and sometimes, even when very ill, doing just a little more than they ever thought was possible. In particular, they can do much for pain and limbs swollen as a result of cancer and its treatment.

Social workers

Social workers are to be found in larger hospitals and working in area teams in the community. Their remit is so wide as to make description difficult but, in particular, they can offer financial assistance, arrange holidays, rehousing, and other help, as well as skilled counselling to enable people to cope with changes in their lives. Unlike the 'paramedicals' just described, they can be contacted directly by patients and families.

Clergy

Last, but certainly not least, of the main groups of professional carers, are the clergy. Throughout the UK there are parishes each with their priests and minister, not only for births, marriages, and deaths, but for the whole spectrum of what has come to be termed 'spiritual care'. By this is meant not religious customs and practice, nor denominational matters, but trying to help with that difficult-to-define part of human nature, the search for meaning. At some time or other most of us wonder and ask such questions as—why does Man suffer? Why do we have pain? Is there a God? Does God care? Why does it matter what I do? What is life all about? All major hospitals and hospices have chaplains and their assistants, always very busy, but they are ever ready to be approached for help.

Of course, not only those of a Christian tradition ask such spiritual questions. These questions are common to Man, whatever his race, colour, faith, or background, hence the importance of keeping in contact with, or making contact with, those best able to help in these vitally important matters whether it is a Jewish teacher or Rabbi, an Islamic teacher, or a local Hindu leader. All have pondered over these issues and can be of inestimable help, each in their own way, each reflecting the traditional teaching or values of their faith.

Sadly, in what has come to be seen as a secularized society, most people from a Christian background have come to see the clergy as representing a denomination, a particular religious group, people who conduct baptisms, marriages, and funerals. This role is important, particularly when the one we are caring for is shortly to die, but no-one should hesitate to make contact with them as *spiritual* advisers whether or not there has been any affiliation to a church or group in the past. If professional advisers do not initiate the subject of how they can help and how they can be contacted there should be no hesitation in the carers or the patients in bringing up the topic.

Cooperating together

It sometimes feels as if we live in an age where everyone has to be an expert and if you are not one you feel useless. This certainly applies in the care of the dying. People not trained as doctors, nurses, or therapists, or anything else which might be deemed relevant, feel as though they have nothing to offer. Nothing could be further from the truth.

The caring relative, whatever his or her other qualifications and qualities, brings something to this work which others may have but in a different form and in much less abundance—love; love which understands and which makes sacrifices, love which tolerates and forgives, love which can do more for someone than anyone will ever be able to measure. It is love developed over years of knowing and usually living with someone, love tried and tested, nurtured through many tears and much laughter, love which has grown in spite of, rather than because of . . .

The person caring brings even more than love. There is their understanding of the patient, his or her likes and dislikes, the fears, the memories, the associations which others know nothing of. They bring the accurate recall of events which might now be important, long-forgotten crises and how the patient came through. Every little insight, every detail from the past, might now help to complete the jigsaw of understanding which can make for better care.

When you meet up with the different professionals, hopefully notebook in hand, be ready to fill them in with information or insights and equally ready to follow their every advice. Some hints which might help are given below.

● Jot down everything as so often suggested in this book. Many misunderstandings, fears, and failures arise because carers forgot what was said. As time passes and they become more tired and strained, they come to remember only what was not said!

- When you do not understand, say so! You will not be unique in not understanding doctors who often admit to being poor communicators and, in any case, they become accustomed to speaking a 'language' or jargon of their own. Just say, 'Sorry, I don't understand that. Would you please try to put it another way?' They usually will!

- It is all too easy, either deliberately or unintentionally, to play off one professional against another. This may be because the different professionals involved all have to explain what is happening and what they are trying to do and, in their attempts to make things comprehensible, they *seem* to be saying different things. A surgeon may speak of 'removing all the tumour I found' yet later an oncologist will be advising how he hopes to treat the cancer left! They are not in conflict or dishonest. The surgeon removed all cancer *visible to the naked eye*, fully appreciating that some cancer cells, invisible except when viewed down a microscope, would still need chemotherapy from the oncologist. The physiotherapist who speaks of helping the patient to walk is not at variance with the doctor who says the patient may never return home. Try to understand and ask for explanations rather than criticizing them.

- Whenever you meet them, whether at home, in a hospital, or a GP surgery, always ask how *you* can help. If they seem to expect more of you than is reasonable, say so. Perhaps they know you better than you think; perhaps you are right and need more help.

- If you feel you need help, then again say so. On no account hold back because it hurts your pride to have to ask. When you are caring for a dying loved one, ultimately only one person matters and that is not you, or your pride, but the one who is dying. Trying to cope without additional help when it is available only leads to more tiredness and sleeplessness, bad temper, and impatience, and finally the patient agreeing to return to hospital or hospice to save you from further hassle and exhaustion. There is nothing to be ashamed of in accepting the offer of a nurse to help at home, a home help to relieve you of some housework, a volunteer car to get you to the hospital, or a Marie Curie night nurse to ensure you get a full night's sleep. This is not defeat. It is common sense.

One final word on this subject. Research has consistently shown that help of the types illustrated here is not always offered to caring relatives. Many carers have no idea what help is available. This is a serious omission because a range of services is available. It makes it all the more important not to sit and wait for them to be offered but to take the initiative and ask about them.

• It may seem unnecessary to say so, but do not go against or do anything without the advice of the professional advisers. Do not give food or drink or drugs the patient is not supposed to have. Do not rest him when he should be exercising, or the opposite. Follow instructions to the letter. If told to give medication every four hours, do so—not four times a day! If told to give a painkiller whether he has pain or not, do so! The doctor really does know best on these occasions. Never hesitate to get confirmation or to say if you do not understand. 'Do you really mean I should give him his medicine at 2 o'clock in the morning?' 'Does he need that pill even if he is not constipated?'

• Though I mention it elsewhere in this book, it is worth repeating here that few things are more trying for busy professionals to cope with than different members of a family all making contact, all asking the same questions, offering their own diagnosis and advice, yet rarely if ever getting together with other members of the family. The correct course is to agree on one or two key people who will liaise with the professionals, speak on behalf of everyone, and convey accurate information back. This applies wherever the patient is, whether at home, in hospital, or in a hospice.

5

Visiting in a hospital or hospice

There must be few places intended to be havens of healing care that are as intimidating as a hospital. Everything seems to militate against relaxed, homely informality. Visiting a patient can become a trial.

There are a number of reasons for this. The layout of buildings and rooms is so different from that at home. Each hospital has its characteristic, distinguishing smell, sometimes of disinfectant, occasionally of urine or a poorly ventilated toilet, but more often than not, a smell impossible to describe which is homely only to the professionals who work and live there. To all this is added the unique cacophony of sounds, the doctors' pagers, the incessant telephone-ringing, the nurse trying to make herself heard by the deaf patient, the trolley carelessly pushed into a door . . . the list is endless. Everyone except the visitor seems to feel at home and know where they are going and what they are doing. Everybody goes about their work without so much as a glance at the bewildered visitor, readily recognizable by the lost look, a bunch of flowers in one hand and a bottle of juice in the other.

That description ignores the deeper reasons why hospital and even hospice visiting can be so difficult and trying. The patient we are speaking of in this book is there because he or she is so ill, in fact mortally ill with a disease which spells death at some time in the near future and probably untold suffering before that happens. Hospitals come to be associated in everyone's mind with hope married to sadness, suffering, and fear. It is at one and the same time the place where hope can be founded, but also where someone will have to say that cure is not possible and every effort must now go into loving, understanding, comprehensive caring—palliative care. No wonder we all have mixed feelings when we go in as patients or as visitors.

Hopefully it is not quite so daunting in a hospice because there so much more effort has gone into creating its homely, relaxed atmosphere and its special ambience. Staff are specially trained in welcoming visitors, time is set aside to speak with them in special rooms, there is less of a 'hospital' feel about it, more flowers, well-planned colour schemes, and the facilities designed exclusively for the comfort of visitors; but even the best hospice is still a special type of hospital, its patients all very ill, all with life-threatening conditions no matter how well and comfortable they may appear with the benefits of such special care.

How can you make visiting easier and possibly happier?

• Get a copy of the booklet each hospital produces, telling of its services, staff, visiting arrangements, and so on, and try to respect the guidelines suggested.

• Never go without that notebook, so often referred to in this book. Go armed with whatever questions you want to put to doctor or ward sister, the answers they asked you for, and all the items of interest you collected to tell the patient.

It's easy to forget how being a hospital patient cuts you off from the outside world. Of course you can see TV or read a paper, but get nothing of the local news and even gossip you would at home. The visitor should go in remembering whom they've seen or spoken to, with news from church, club, neighbours, and enquirers.

• Do not expect the patient to entertain you, like a gracious host. If he wants to talk about how he feels or what has happened to him since your last visit, rather than listen to your news, so be it. Show interest in how he is and the news he wants to share, then give him a rest as you recount your tale. Nothing is more depressing for a very ill patient than to be asked clinical questions by each well-meaning visitor! How well I remember a patient's wife standing in a doorway of his ward, her first loud-spoken question to him heard by everyone else in the room, 'Well, have they moved yet?' Everyone knew that she was referring to his bowels!

• While not denying the 'down' side of his report to you, try to focus on the 'up' side, the progress, the pain which is lessening, the appetite returning, the better sleep, those few faltering steps, the questions which he has had answered. Always be positive without being unrealistic or overbearing. At all costs, resist the temptation to urge him to 'pull himself together' or 'cheer up'!

• Resist the temptation to 'sympathize with the patient', to feel how he or she is feeling. It's doomed to depress or annoy and is, in any case, very insensitive and patronizing. How can anyone sit by the bedside of a dying person and say, 'I know just how you feel'? There are others who delight in capping the patient's every story with an account of their own last spell in hospital, their operation, or even their unsolicited, often damning opinion of the specialist looking after the patient. Quite unforgivable is showing one's own operation scar to the patient you are visiting—but this does happen!

• When the patient says things which upset you or asks questions which you cannot or do not feel you should answer, make a point of speaking to the ward sister or house doctor as soon as possible and getting their advice. This is dealt with in more detail in Chapter 11.

• Give the patient your undivided attention. It is hurtful or irritating to lie in bed and watch your visitors eyeing the pretty nurses or attempting to glimpse the TV, or even worse (but so often seen) reading your paper as he eats the grapes.

  Having said that, it has to be admitted that it can feel a long vigil at the bedside of someone ill and near to death when conversation is impossible. Here it is quite in order to sit quietly as the life of the ward or room goes on around you, possibly reading or even dozing, provided that when the patient is awake he is given undivided love and attention.

• Do not smoke immediately before sitting at someone's bedside—it can be very upsetting. The patient who used to enjoy a smoke will yearn for one, while others will probably feel nauseated by the smoker's breath and clothing. In the same way, common sense dictates that you should not drink alcohol or have a strongly flavoured meal such as curry before visiting.

- Again, common sense would suggest that caring visitors will not give the patient any food or drink (and certainly not alcohol) without prior discussion with, and consent from, the ward sister. It never ceases to surprise doctors and nurses how often this happens. Some people even bring in pills and potions without any permission or encouragement to do so!

- The doctors and nurses, especially the ward sister or staff nurse, are there to help by answering questions. Nevertheless they do have much else to do, even during visiting times, many other patients to look after, many other relatives and friends who want to speak to them. This is so often forgotten. Some visitors expect them to drop everything to speak to them, ask to see them at every visiting time, talk to them for far longer than they need to, and by so doing upset the work of the ward and deprive other visitors of their help.

The answer is simple. If the ward has a policy of setting apart time for visitors to meet the doctor or sister, abide by that time and use it well with the help of that notebook. Otherwise, ask to see the sister at her convenience during visiting times and take up as little time as possible or telephone to make an appointment to see her or one of the doctors. Always go prepared for serious talk, time well spent and not just for a chat.

The doctor who is on the ward for most of the time (and in fact so much so that it often looks as though he or she never goes to bed) is the 'house doctor', well able to answer most questions except for those needing specialist knowledge and information. In that case, you would be advised to get an appointment with the consultant or he may seek you out. He is the senior doctor, the specialist, with ultimate responsibility for the patient's medical care.

Practices differ greatly in different hospitals. In some, the principal relatives are sought out and everything explained to them. In most hospices this will be the case, but in other equally good units the responsibility for making an appointment rests with the visitors. One final point—when there is a large family it is clearly impracticable for each member to see the nurse or doctor. They should decide among themselves who is best to do this, then convey all that was said to the rest. It is remarkable how many people tell the doctor that they do not speak to other members of the family or, even when they do, they do not trust them to pass on information! Doctors and nurses are always happy to support families in their stress and sadness but have more to do than referee family feuds.

- Do not expect junior nurses or nursing auxiliaries to update you or answer questions. They are not qualified to do so. Any of them can direct you to a qualified nurse able to help.

- If the patient says something at odds with the information given by the professional staff, remember that the patient may be muddled or frankly confused though very convincing in what he says. Relatives often come to nurses expressing horror that a patient has told them he is going home the next day, when the relatives know he is very ill. Usually the true situation is that no-one has even thought about him going home, and if they had the family would have been invited to join the care team and discuss the many implications. However, at other times the patient may have a genuine grievance about his care; and in the above example he may be telling you that he wishes he could go home.

- If children want to visit a hospital or hospice, check first with the sister. Always ask if there is a playroom or creche where the children can be looked after while the parents spend longer with the patient. Children often get bored and restless at the bedside. It may help if you take a book or toy for them to play with.

Visiting needs to be planned—just like every other aspect of care. Thoughtless visiting does not help the patient and can leave him weary and sad. Well-planned visiting can be a happy experience for all concerned, even when the patient is so ill.

6

The truth, the whole truth . . .

It's probably not an exaggeration to say that nothing is more important, nor anything more frequently talked about when caring for the dying, than how much the patient knows and should be told of his condition. Everyone seems to have a different opinion though, it must be said, the patient's own view is often never sought. Everyone seems to know what is best for the patient without asking him.

The 'man in the street' will probably say that when his time comes he wants to be told the truth, the whole truth, and nothing but the truth. He wants no lies, no false hopes, no whispering behind his back by his family and doctor. Very sensibly and understandably he will point out that it is *his* life, *his* future, *his* business, and *his* right to honesty, no matter how painful the facts.

So far so good. When someone very near and dear to that same person is found to have a mortal illness, his views are likely to be quite different. He will urge that the truth be withheld so that hope can be maintained. He will probably do all in his power to keep his relative in the dark. He will make up the most implausible explanations to protect the patient, telling him everything except the truth. He may try to persuade the doctors to do the same, sometimes going as far as to try to put an embargo on anyone attempting to explain the true state of affairs.

Until a few years ago, he might have found allies among the doctors. It isn't long ago that a survey of doctors found that 80 per cent of them felt that the awful truth should be kept from patients unless they demanded to know—but when the same doctors were asked if, when *their* time came, they would want the truth withheld from them, 80 per cent said no! Times have changed and now most doctors, particularly those with considerable experience in this field, would say that patients actually benefit from a policy of sensitive truth-telling and that it makes their care and subsequent comfort so much easier. In any case, it can be strongly argued that everyone (whatever their state of health or illness) has a *right* to honesty and a mutually trusting, respectful relationship with their doctors.

Many factors have led to this change, not least a deepening understanding arising out of what patients have taught us. Time after time they have been thought to be completely ignorant of what ails them, only to reveal that they have known it for a long time. Even when not actually 'told', they deduced it and then had to keep their insight and knowledge secret from the family or friends all conspiring to protect them. How, then, do people deduce the diagnosis and its significance when so many people are trying to shield them from it?

They start by suspecting something is seriously wrong and soon have confirmation. As people get older, inevitably they have friends and relatives unwell or dying. They learn from that experience how serious illness strikes, what tests doctors carry out, what treatment is offered. It is natural that they should see parallels in their own illness. Some have family histories of cancer or heart disease. Suspicions grow stronger the longer it takes the doctors to make the diagnosis, the assumption being that if it was simple and straightforward it would not take all these sophisticated investigations. They suspect it when they are given vague explanations, when doctors clearly find it difficult to explain things. They cannot help but notice the embarrassment and unease on the faces of doctors young and old, and sense the palpable tension when close relatives are spoken to, and when relatives so unexpectedly come home from all corners of the world. Others learn of the diagnosis from furtive glances at case notes left within their reach, or from tutorials being given on their condition to medical students. Time after time patients recount to doctors, including me, how they have known almost since they first went to their doctor but kept up the pretence of ignorance, not so much for their own sake but for the sake of others, particularly the family—but also their doctor!

In spite of this, most people will continue to claim that *their* relative is different and really does not know. They describe how they have always been close and had no secrets, nor indeed ever wanted any. If he or she had known, or even wondered, they would have spoken about it, but that has never happened. Clearly the patient does not know and does not even suspect! What is overlooked is that the nearer people get to the end of life the more considerate and thoughtful they often become. Even the most open and truthful person will tell and live a lie, just as the relatives are now doing, to save someone they love from pain. It is also forgotten that most of us do not readily bare our souls to all and sundry. We instinctively select a confidante, sometimes the most unlikely person and often not a family member, and no-one else is made privy to our secrets. In hospital a patient may act as though he genuinely believes he only has a chest infection when he is speaking to doctors and qualified nurses, yet he will talk to a porter or nursing auxiliary or the ward maid, telling them that his days are numbered and that he knows he will die shortly. To the end of life he may continue with this pretence with his family. Perhaps a true story will illustrate it.

Some years ago I was looking after a man with lung cancer, by then in the final weeks of his life. For two years he had been failing, losing weight and his vitality, and now was almost completely bedbound and able to do very little for himself. His wife proudly boasted how she had kept the truth from him and was confident that he had never suspected anything was seriously wrong. In fact he was still planning a long touring holiday in Spain the following year. No-one was to 'tell him'. One day he asked to speak to me and produced a list of no less than 22 questions. They ranged from how many days of life were left to him to exactly how he would die and detailed advance planning of his own funeral service! He had just finished telling me how grateful he was for the honest explanations and for all the information that I had given him and was discussing the hymns for his funeral service when his wife appeared at the bedside. Without a pause to draw breath, he excitedly told his wife what a help I had been, describing the best car to take to Spain, the preferred routes and hotels, all he and his wife needed to know to make next year's holiday one to remember! To his death the following week, he never spoke of his insight to his wife and she remained convinced that she had protected him— until of course his papers were read with the carefully planned funeral details. Far from being unusual or unique, this type of story could be repeated many, many times. *Most people know what is wrong with them and how serious it is.* We only fool ourselves and we do them no good if we deny them the information they need, particularly if we deny them any opportunity to share their feelings about it with those of us whom they love.

Clearly there are several possible reasons for relatives wanting to keep the truth from each other. They want to shield them from pain and distress, as a mother would with her children. They also want to protect themselves, not knowing exactly how the patient will react to the knowledge nor how they will cope with the patient's reaction. They are also fearful about their own reaction and whether or not they will cope when there is no longer any veil of secrecy between them. Many wrongly assume that if the patient is told he will not stop crying or talking about death. They are wrong —most dying people do not behave like this, particularly if they are well cared for in an atmosphere of openness and love.

It is helpful and interesting to remind ourselves that even professional health workers have the same feelings and fears on this subject. Studies of students and young doctors have shown the same reasons for withholding the truth, or at least most of the truth—fear of the patient's reaction and fears of being unable to help them, however they react, when they are 'told'.

It should by now be clear that what has been termed the 'conspiracy of silence' cannot be, nor should be, sustained. No matter how well intentioned, it is not helpful. In fact it may be positively detrimental and counter-productive.

If a doctor, whether of his own accord or prompted by a relative, elects to tell a half-truth to someone with a fatal illness, who then can the patient trust? To whom can he turn when he wants honest answers? Can he trust that doctor in any other aspect of his care if he cannot trust him to be honest about the most important time in life? Some years ago I met a woman with advanced cancer. She described how only a few years before, when her mother was dying of cancer, she had forbidden the family doctor to tell her mother the truth. Instead she plied him with the untruthful explanations she thought would shield her mother. Now, faced with her own terminal illness, she was terrified—not sure what was happening, unsure whom she could trust, lonely and frightened. 'How can I possibly trust my doctor? He lies so plausibly. I know, because he did it for me when mother was dying and now I simply can't believe him or trust him! What I need now is honesty, truthful facts, explanations, safety—I don't want lies!'

When someone's life is under threat, they need the strongest bonds of love with family and friends, not doubts or suspicions, and certainly not the necessity to act out a lie, something which they have tried to avoid in their lives together.

People wrongly assume that 'openness and honesty' means hours of agonizing conversations about dying, death, funerals, and rituals. Nothing could be further from the truth. If there are no secrets and many shared feelings, there is little call for discussion and interminable repetition. Tensions are less, conversation is freer, silent companionship more natural.

Let us look at some frequently asked questions about such a caring, open relationship.

Q. What do I do if he wants to talk about dying?

A. Allow him to do so but do not assume he wants answers to his questions. Do not try to change the subject or cheer him up. Do not be afraid to share your feelings—your sadness, your frustration, or your disappointment at how things have turned out. If he asks medical questions which you cannot answer make a note of them to put to the doctor.

Q. What if he *stops* talking about his illness and what is happening?

A. Leave him to his thoughts. In a loving relationship we do not need to be talking all the time. It is a sign of a healthy relationship when silence is both permitted and peaceful. You both need quiet times to ponder and prepare.

Q. Should I talk about my feelings?

A. Yes, but be exceptionally considerate. Choose the time, watch his reaction, and do not expect him to carry your emotional burden as well as his own. Whether a partner or friend, never shy away from reaffirming love and respect and from reminding him what your relationship means (and therefore by implication what you will miss).

Q. What if he jokes about what's happening?

A. Humour can be a release from tension, an alternative to tears yet often so close to them. Don't be surprised if the humour is sometimes 'black humour', often making fun of his state and what lies ahead. This is normal for the patient but may sound insensitive or distasteful to others, but make no comment. Many couples recount how they laughed and cried together more in the final months and weeks than in the previous years together. If a remark meant as a joke hurts you then say so, so that he may remember how fragile you are and that both of you are suffering. For example, many women tease their husbands about finding another wife when they have gone. They may be actually hoping their husband will eventually remarry, but the idea may upset him when she expresses it.

Q. What if he talks to other people and doesn't talk to me?

A. This is very likely to happen. He is protecting you. Perhaps it is easier to talk to others he is not so close to. Perhaps he does not know how you will take it. If it worries you or upsets you, then ask if he finds it difficult to talk about something. Ask yourself how easy you found it to talk together about different things *before* he was ill.

There will be many, many occasions when you will wonder if this policy of truthfulness and openness is right for you. You will sense his fear, see his apprehension, and ask yourself time after time if it would not have been better if you or 'they' had not told him. Every time he is quiet or depressed, you will want to blame the doctor for telling him. Then, on other days, you will have no doubt. You will feel the warmth of trust and respect, the comfort of knowing you are not living a lie or unintentionally hurting someone you love. You will know the indescribable joy of silent companionship with him as the end approaches and no more words are needed.

Some may say that truth hurts. Those who care for the dying know that lies destroy but honesty heals every wound.

One final thing remains to be said, self-evident as it may seem to some readers. Each of us can absorb only so much information at any one time. When it is good news we can bear to hear a lot of it: when it is bad news, we can only take a little before a 'safety curtain' falls and little more is remembered. Too often the inexperienced doctor or nurse, with the best of intentions, says too much or explains in too much detail—often, it must be said, under pressure to do so from relatives and carers. The result can be worse than if nothing had been said. The most important facts are forgotten and the more trivial details are remembered. Each person, patient or carer, should have information given at the time and in the detail when it seems right *for them*, no matter what others may say or believe. It is certainly wrong to withhold vital information. It can be equally wrong and damaging to try to give too much at the wrong time or in the wrong way.

7

Food and drink

From childhood we are brought up to believe that certain foods do us good and some are harmful. The trouble is that the evidence on which these hallowed beliefs are founded is always changing. What was recommended yesterday is today being blamed for some illnesses! Even when the expert opinion is consistent, it can still be difficult to understand. For example, we are told that milk is good because it contains so many of the nutrients we need. That is true but then someone reminds us that fat in the form of cream has some disadvantages and, no matter how nourishing milk is, it is always constipating. This is one of the problems with so-called 'invalid' foods, largely made from milk and dairy products. They are easily swallowed and digested but may make constipation worse. We can all think of many other examples. Dietetics is a science in itself, very necessary but sometimes baffling to the untrained people who only wants to do the best for the patient.

The dilemma is made worse by the fact that when somebody is very ill, he can usually take only small helpings of anything and what he now fancies may be quite different from what he has always liked in the past. This can be puzzling and challenging for the carers. The person who would never think of eating porridge now says that it is all he or she wants and proceeds to ask for it several times each day. The one who has never liked ice-cream now finds it the easiest thing to eat and cannot get enough of it. To complicate matters even further, many very ill people cannot decide what to have and when offered something special which only a few hours before sounded attractive, they send it back. Carers go out and buy expensive treats like Dover sole, only to have it rejected, untouched.

Of course there *are* some conditions which require special diets, strict adherence to which is said to be essential for their well-being. Examples might be the low-protein diet for those with kidney failure or the strictly controlled sugar intake for the diabetic. Such patients need both the skilled guidance of dietitians and the patient understanding and cooperation of the carers.

The final challenge comes when the illness is so advanced that the patient cannot, or will not, keep to such a diet or will take so little of anything offered that it appears they might starve to death. In our western society we have been brought up to believe that our health and survival depend very largely on a good, balanced food intake. Inevitably, when that intake gets progressively less and less, the carers fear that the patient's decline is a direct result of his poor intake. Not only that, but in our society eating has become a social function, a time to sit together and enjoy each other's company and chat. A well-prepared, much appreciated meal is a hallmark of a caring relative, particularly a wife. Often it is all she feels she can do to keep her husband well and near her. A refusal of food looks like a refusal of her or a reflection on her cooking or caring, a rejection of all she represents.

In short, a diminishing appetite or changing craving for new and different foods bewilders and upsets most carers. They see it as a personal failure and, particularly with the dying, come to blame themselves for the speed of his decline. They feel there is very little they can do to help, particularly when the patient is suffering from something as complex and bewildering as cancer, but at least they can, as they have so often in the past, prepare appetizing, attractive meals. When that fails they feel *they* have failed and blame themselves for what is happening to the patient, or the doctor for not prescribing an appetite stimulant.

Simple principles

Without in any way diminishing the important contribution dietitians can and do make to the care of the sick, it is reassuring to be told that the principles are simple and putting them into practice well within the power of the untrained carer. Let us look at them.

• Loss of appetite (anorexia is the medical term) is almost inevitable at some time in the course of any illness and particularly so in the later stages of it. It is not the result of someone's failure or inadequacy. In a few patients the doctor may be able to stimulate appetite with steroids but the effect is often short-lived and in many patients steroids are contraindicated.

• Loss of weight and energy (cachexia and asthenia are the medical terms) is an inevitable feature of almost all cancers (except those originating in the brain) and also of many life-threatening conditions of the heart, lungs, and kidneys to name but a few. What must also be emphasized, because very few people either understand or accept this, is that changes in dietary content or quantity make little or no difference to this wasting in the late stages of the illness. Put simply, there comes a time when even the most nutritious, most attractively prepared and presented food will not stop the patient losing weight. What is far more likely to happen is that if he tries to eat it when his body cannot digest and utilize food, it will upset him, making him feel sick or guilty that he is disappointing his carers.

• Very ill people, particularly with cancer, never seem to have two days alike. This applies to their energy and what they can do and, equally, to their interest in food. What they like today, they may be unable or unwilling to try the next day. Just when the carer feels that she or he has at least found something nutritious and acceptable, the patient refuses it. This can be very frustrating but is always worth discussing with the doctor or visiting nurse.

• There comes a time when the best dietary advice, and probably the only advice needed, is to give the patient what he wants, how he wants it, when he wants it, provided the helpings are small. What does this mean in practice? It means that the only plate needed is a small side plate and never a dinner plate; that standard meal times and what is traditionally eaten at those times are ignored; that the patient gets what he fancies every few hours, remembering the curious fact that most people coming to the end of life eat more at breakfast than at the middle of the day, and often nothing in the evening.

• In the last weeks and days, the interest in food may disappear completely and nothing whatsoever will stimulate it. This is nothing to worry about and is, in effect, the body's way of saying it cannot digest anything. When it comes to food, the body seems to know best. What the body does *not* know enough about is the amount of fluid it needs. Many people fail to drink enough even when well able to, not appreciating how a good intake of fluid (provided they are able to produce and excrete urine) keeps them brighter, staves off confusion, and goes some way to lessening constipation. Every effort has to be made to encourage the frail patient to drink as much as possible for as long as they are able to. The secret is to offer a variety of flavours, *preferably as cold as possible*. The nearer a person is to the end of life, the more they will prefer and appreciate iced drinks. On this occasion the old wives' tale is wrong—sipped slowly through a straw, ice-cold drinks do not produce a 'chill in the stomach'.

Useful hints

- Offer very small helpings of different items, every few hours, no matter how unusual they might sound for that time of day. For example, someone might manage some porridge or a bran cereal at 8 a.m., some cold stewed fruit at 10 a.m., and a lunch of a single course at noon—perhaps clear soup or mashed potatoes and gravy. At 2 p.m. they might like a few spoonfuls of ice-cream or yoghurt, at 4 p.m. a little more fruit or some custard, and perhaps little else that day, or a small serving of fish or chicken in the evening.

- Keep a simple diary of what was offered and appreciated (or returned untouched).

- Keep a supply of cold drinks in the refrigerator, taking out just enough of what the patient wants to have beside him. The range should include lemonade, tonic water, fruit juices, or the old favourite of so many athletes—a mixture of equal parts of skimmed milk and soda water, a most refreshing drink for the very ill.

- Also kept chilled in the refrigerator can be a small dish of thinly sliced tinned pineapple, each slice no bigger than a postage stamp. To make them even more attractive, they can be sprinkled with icing sugar before frosting. The patient can suck them between meals and then, if they wish, discard the pulp. Pineapple both stimulates saliva and the digestive juices and also has a chemical effect on the mouth which aids good mouth hygiene. Fresh pineapple has the same effect but can often taste too sour or be astringent, particularly if there is any infection such as thrush (referred to as candida by the doctors and nurses).

• If possible avoid a diet composed exclusively of commercial 'invalid' foods, nutritious as they are. They are usually constipating and very soon become boring.

• When solids are refused, do not despair but do everything possible to encourage as large a fluid intake as possible. If the colour of urine excreted is clear or pale, the patient's fluid intake is adequate. If it is dark it means the urine is concentrated, usually because the intake is insufficient, or that the patient has jaundice—something which will be obvious anyway.

• Remember how enticing is well-presented food. Pour a few drops of coloured syrup or liqueur over ice-cream, put a sprig of parsley on fish, offer jam or lemon curd on rice pudding—anything to make it look (and possibly taste) more attractive.

• Try to protect the patient from kitchen odours no matter how appetizing they may be to others. They are almost universally reported by the critically ill as upsetting or nauseating. Unless the patient asks you to, do not eat your meals with him or her. It can be upsetting to see someone eating larger helpings, particularly if it is food they always enjoyed.

Parenteral feeding and rehydration

In simple English this means putting food and fluids into a patient by a route other than the mouth. In practice is usually means using tubes inserted into the bloodstream or via a surgically constructed hole in the abdominal wall (percutaneous gastrostomy).

There are many occasions when this route is not only useful and convenient but probably the only way to feed someone, and is life-saving. Examples are after certain operations, when patients are unconscious, and when, for one reason or another, life itself depends on this intake. This is *very rarely* the case with far-advanced illness from which the patient is going to die. On these occasions it is not the lack of nourishment or fluid which is threatening life but the underlying disease process. As already explained, forcing food and fluid into someone nearing the end of life from cancer, AIDS, and many other illnesses is unnecessary, unhelpful, and usually very distressing, first to the patient and finally to the carers who feel that the drips and tubes and equipment have become barriers between them and the one they love.

It cannot be stated strongly enough that such measures rarely if ever lengthen life or improve the quality of life for the dying person, much as they are often asked for (or even demanded) by so many relatives and friends who convince themselves that the patient is dying of starvation or dehydration. No patient ever complains of dehydration. What they do mention is a hot, dry mouth, a sense of the tongue being swollen, difficulty in speaking, a bad taste in the mouth, and sometimes a sense of bad breath, all of which can so easily be prevented and treated by sucking chips of ice, frosted pineapple slices, and chips of effervescent vitamin C.

8

Planning care at home

Every aspect of the care of a dying person needs thought and planning. Nothing should be left to chance. Everything which can be anticipated should be thought of and preparations for it made. In this way, care can be improved and, who knows, possibly more people will achieve their wish to remain in their own homes. Planning where and how we look after the patient at home is our new challenge.

Some things may seem self-evident but are nevertheless worth repeating.

Motivation

A willingness to try to care for someone at home is infinitely more important than having ideal accommodation. Similarly, a genuinely loving atmosphere is more than a substitute for one characterized only by professional skill and expertise. If you really want to care for someone at home, there is usually a way to make it possible—if not until the end, at least for much longer than anyone would have thought.

Teamwork

Just as the staff in good hospitals work as teams, so should carers at home. Such care cannot be given by one person, nor by a group of relatives or friends who do not work as a team.

Perhaps it is worth reminding ourselves what we mean by a team. It is a group of people with different skills and strengths, working towards a common goal, under a designated leader. That leader is not necessarily the most skilled, experienced, or the eldest, but someone elected to the role because they can coordinate, lead, support, encourage—and sometimes rebuke—all to help the team achieve its goal.

In home care of the dying, the team (though they may not have thought of themselves as such) are the relatives and friends who have all agreed that everything should be done to nurse him or her at home and have committed themselves to this goal. It might also be argued that the patient should be in the team too. They bring different skills, though some may feel they have nothing to offer, but all share a love and concern for the patient. They should all be prepared to give of their time, energy, and emotions. They care so much for the patient that nothing feels like a sacrifice. Usually without realizing it, one emerges as the leader, the organizer. It may be the husband or wife, an elder brother, or a lifetime partner but, whoever it is, the others instinctively listen to him or her and take the lead from them.

Organization

Like any effective team, the carers must organize themselves. There has to be a duty, 'on-call' roster. While one works, another remains in the house. One sleeps, another keeps vigil. There has to be allocation of duties. One does shopping, another tidies the sickroom, while another does something else. Each task is important and there need be no jealousy or resentment about one doing more than another. A team works towards a common goal.

Home care so often falls down because of lack of organization. One person attempts to be all things to all men, or a group of carers duplicate each other's work or squabble as their grief and frustration increase. Finally, the patient is almost forgotten in the unnecessary disorder or sickening fatigue everyone experiences.

Personal cost

Carers must talk things over together once in a while. They need to review what they are doing, how they are doing it, whether it is effective, how tired they are, what changes might help, and, so very important, how each one is coping. There is a saying among doctors and nurses caring for the terminally ill that dying can bring out the best in the patient and the worst in the relatives! The strain and grief seem to highlight differences and weaknesses, bringing back to mind old feuds and, tragically, diminishing what could be very effective and adequate care.

Simply sitting down together with a mug of coffee, each person in the house being honest about what they feel in themselves as well as what they feel about others, resolves most tensions and brings a renewed appreciation of how much they all have in common. No-one would deny that such 'talk it through' sessions are not easy or comfortable. Why not invite the family doctor to join in or chair the meeting? It can make all the difference.

Many times in this book mention is made of anger, because it is such a common but destructive emotion. There is nothing wrong with feeling angry, nor with expression of anger, *provided* that attempts are made to learn from it and prevent it damaging people and relationships. This is particularly the case within families. One member is angry because he feels he is left to do everything, another is angry because 'they' (usually meaning the doctors) are not doing enough, or doing too much! Another is angry for no apparent reason but it looks to onlookers as though he is guilty or embarrassed but not big enough to speak about it. For each it is a feature of their loss and grief. They are trying to cope by blaming and lashing out at those nearest to them. Unless this is tackled by all concerned, it soon heralds the end of home care for the dying and leaves everybody with a bitterly unhappy memory of that time.

Anticipation

Sit down with the family doctor (or palliative medicine specialist if one has been called in) and try to get the picture of what will be entailed in home caring. Ask about what the patient will experience, what pains he might develop, what crises there might be, and how each would be dealt with. Discuss nursing equipment, special laundry requirements, diet, bedding, and exactly how much nursing will be needed. Do not hesitate to ask how much he or she will require special skills and how much might reasonably be done by the untrained person. Be honest enough as an individual or a family to express fears and misgivings rather than promising a quality of care you cannot hope to provide.

One thing cannot be said often enough. *Most people fear the unknown more than the known.* This applies to the patient. It also applies to all the carers. It can certainly be daunting when the doctor lists or graphically describes what may lie ahead, but after that it begins to lose its fears. You now know what to expect and can prepare for it, no matter how apprehensive or ill-qualified you feel.

Reviewing

Keep reviewing the situation. An initial pledge to keep a relative or loved one at home does not make it a personal failure if he or she has later to be admitted to a hospital or hospice. It is often said that most dying people would prefer to die at home but that assertion must be questioned. It sometimes means they would like to spend as long as possible at home but not necessarily to die at home. Do not make a promise you cannot keep—'I've promised her she will never again go back into hospital'. Aim to care and cope at home for as long as possible, always putting patient before self. If the time comes that admission to hospital or hospice is the right thing, you will not see it as failure but rather as a need for more pairs of hands to help.

Planning the sickroom

Plan the layout of the sickroom to make it as homely yet as efficient as possible. Just because the bed has always been in that position and the chair in that corner doesn't mean they must stay that way. What are the principles which dictate what is needed and where?

The bed and essential furniture

The bed should preferably be a single one to enable helpers to reach the patient from either side. It should now be placed near the centre of the room with space all around it for the attendants. This is usually possible even in the smallest bedroom if unnecessary things are taken out. If possible it should face the window. On one side should be a simple bedside table for drinks, tissues, and mouthwashes. On the other side there should, if possible, be another small table on which can be a light to shine over the patient's shoulder but, if needs be, everything can be on one table. If there is any difficulty whatsoever in the patient getting to the toilet, a commode should be placed in the room where the patient can reach it. Lastly, there should be a comfortable armchair, preferably facing the window, for the patient or relatives and friends sitting with him. If funds reach to a reclining chair, so much the better.

Bedding and aids

Bedding should be light and washable, with as many pillows as possible. The frailer a person is the more they appreciate being supported, day or night, by soft pillows or cushions. Community nurses can provide polythene sheeting to put on the bed under a drawsheet if the patient perspires a lot or is incontinent. The doctor can prescribe incontinence pads and community equipment stores have available urinals for men, bedpans, commodes, backrests, head-rests and neck-rests, footstools, and real or synthetic sheepskin pads for elbows, buttocks, knees, and heels to prevent pressure sores. In many places they also have on free loan beds similar to those used in hospitals, making the nursing of a frail, bedbound person so much easier for all concerned. The doctor or nurse may be able to put the carers in touch with a special health service laundry for soiled bed linen.

Cutlery and crockery

The patient who has difficulty using normal crockery may feel much happier with special feeding cups, often with one or two large handles, and a small mouthpiece or a hole through which a straw can be put. When people with old or arthritic hands cannot use normal cutlery, one can either borrow or buy special knives, forks, and spoons with padded handles, and plates with special edges which prevent spilling of food. The patient who needs a liquidized diet can either buy a liquidizer or, on occasion, get a modest grant to buy one. Better still, you could ask a neighbour who has one if it can be borrowed.

Waste disposal

Inevitably people worry about waste disposal—dressings changed by the visiting nurse, used disposable syringes, old needles, and empty injection vials and ampoules. Advice will always be given by the visiting nurse but basically it is that everything should be wrapped up in old newspapers so that needles and glass cannot injure anyone and then placed in a bag specially provided. When many needles and syringes are being used, a special 'sharps box' will be provided by nurse and, when full, taken away by her and replaced with an empty one. When her instructions are followed to the letter *there is no danger whatsoever of infection or injury.*

When speaking of infection, one must mention what happens when looking after someone with AIDS. Once again, clear guidance will be given to the carers and, it must be emphasized most strongly, *there is then no risk of infection for the carers.* After the death, the undertaker need not be told that the patient died of AIDS but merely that strict 'infection control' procedures are required. The body is then put in a special bag before being placed in the coffin.

Equipment disposal

After any death at home there remains the task of disposing of equipment and, in particular, any medicines, syringes, needles, and controlled drugs (morphine, heroin, and the like). Often a substantial store has built up and the advice of a doctor and nurse is needed.

Borrowed equipment can be taken back to the stores (or collected by arrangement). The doctor will probably supervise the disposal of all pills and medicines except the controlled drugs. These he will take away and record their disposal in his records. He may suggest that others are taken to the local chemist who has authority to collect and dispose of them. To many this might seem a waste when they might be of use to others but they can rarely be used for other patients because of pharmaceutical problems of storage, age, and shelf-life.

Noise

A sickroom does not have to be silent but it must be possible for it to be quiet when the patient needs it. A TV for the patient should have a remote control and the TV for the family kept as quiet as possible. Even the children who normally make their own inimitable din can usually be kept quieter when they know it is for the sake of daddy or grandma upstairs. Do not attempt to shut out all the normal noises of the world outside—they actually help the patient who does not want to be cut off from the world with all its memories.

Ventilation

Good ventilation is very important. There is no harm in a window being left open so long as the patient is warm and comfortable. A stuffy room is unpleasant and soon makes a frail patient sleepy.

Obnoxious odours are common, particularly with some cancers, and can be removed with special deodorants, but one should not use air-freshener aerosols which later will always come to be associated with the patient's death. On *no* account should any visitor smoke in the room even if the patient smokes.

9

Worrying events

It has to be said that many, possibly most, people with a fatal illness gradually get frailer in their final weeks and months and eventually die very peacefully without any event which could be called a crisis. This is not to say that their care is not a challenge to the family and friends, but they can often look back on that time and be relieved that nothing unexpected or catastrophic happened which alarmed them. This chapter looks at some of the things which can and do sometimes happen, events which are worrying or even alarming and can often make home care difficult or even hospital visiting fraught.

Urinary incontinence

Many things can cause a person to lose control over their bladder emptying, only a few of which are diseases of the bladder itself. There may be a slight infection, a nearby part of the rectum or bowel which is loaded with hard, constipated motions, or simply such weakness that the muscles which act as a valve cannot work properly. In older men there may be an enlargement of the prostate gland under the bladder or the cause may be as simple as the patient being so deeply asleep or sedated that they do not realize that they need to pass urine.

Whatever the reason the doctor will always try to identify the cause and treat it in one way or another, appreciating how embarrassing it is to a patient and how much extra work and worry it causes the carers. A specimen of urine will be taken to examine for infection and to identify the most appropriate antibiotic. The rectum ('back passage') will be examined to exclude constipation or an enlarged prostate, and the nervous system examined in case the nerves which control emptying are not functioning properly. If the cause is simply that the person is sleeping too deeply at night, any sedatives might be reduced. Rarely will the doctor suggest reducing fluid intake, because adequate fluids are so important.

Occasionally the patient will need to be catheterized. This simply means that a special rubber tube will be introduced into the bladder via the urethra after numbing it with an anaesthetic jelly, and a small balloon is then inflated near its tip to prevent it falling out. The urine may then be emptied every few hours or allowed to drain freely into a special bag at the bedside or, if the patient is up and about, one strapped to the thigh so as not to be visible to others. Every few days the catheter is flushed with a special cleaning solution, often by the nurse, but this is a skill easily learned by a relative if they would like to help in this way. This helps to prevent infection (an uncommon problem when a catheter is introduced and looked after correctly) and reduce the cloudy deposit which can accumulate in the bladder.

Often the catheter can later be removed, but occasionally it must remain in and be changed every month or two. Urinary incontinence should rarely if ever be a cause for someone having to be transferred to hospital, particularly if there is a special home laundry service available for any wet sheets.

If the patient develops acute urinary retention, the doctor will again define the cause and in the very seriously ill probably insert a catheter and leave it in as described above.

Faecal incontinence

Losing total or even partial control over one's bowels is always deeply embarrassing for the patient and a cause of such distress and extra work for families that when it happens frequently at home such patients often have to be admitted to hospital.

Again, there are many causes but the commonest can be helped. When someone is frail but badly constipated, the hard stools cannot be pushed out. The result is that they build up in the rectum getting harder all the time and fluid passes over them eventually leaking out like diarrhoea. If this cause is not diagnosed and people continue to think the patient has genuine diarrhoea, he is often given pills or medicines to stop diarrhoea with the result that the constipation is made even worse! So common is this picture that doctors have a saying, 'The commonest cause of diarrhoea is actually constipation'.

Simply by skilled emptying of the bowel and the use of something to keep motions soft and easily expelled, the leak can be stopped. The problem is that so many people continue to believe that because someone is not eating much, there will be nothing to get rid of, which is totally untrue. A reduced intake and one mainly composed of milky, invalid foods increases constipation and produces less bulky motions, thus making them difficult to expel.

Another cause, the exact opposite of that described above, is the over-enthusiastic use of laxatives or purgatives such that the patient cannot get to the toilet or commode in time, or gets little or no warning of the need to do so.

When the patient has a cancer in the rectum or nearby organs there may be a leak of blood and mucus (a clear jelly-like fluid) over which there is no control. As with urinary incontinence, the cause may be the overall frailty and muscle-wasting of the patient.

It is understandable that this problem can herald the end of care at home but disappointing that such causes as constipation or impaction (the lower bowel blocked solid with hardened faeces) should have been allowed to develop. Sometimes the fault is the doctor's in failing to prescribe an appropriate laxative when other essential drugs are well known to be constipating. At the other times the fault is that of the patient or carers who regard bowel action as so automatic that no attention to it is needed, or when the patient says he has so many medicines and pills to take and decides to leave off the all-essential laxative ('After all, how can he be constipated when he's not eating?').

The need to observe the motions and keep a record of bowel actions at home cannot be over-stressed. It can make the difference between staying at home or going back to hospital.

Bleeding

Everyone is worried by bleeding whatever the cause, wherever it happens. There is nothing to be ashamed of in being alarmed by it. It must always be reported to the doctor who will find and explain the cause though, it must be said, he may not always be able to prevent it happening again.

Bleeding from the skin is always controllable whatever the cause, be it a bedsore or an ulcer. When it shows in phlegm or spittle (haemoptysis) it may be due to something as simple as a mouth ulcer or a chest infection or it may come from a lung cancer in which case radiotherapy will help. When blood appears in the motions it is most usually caused by constipation and piles (haemorrhoids) but a cause must be found. Blood in the urine (haematuria) is usually from the bladder but can be from the kidneys, particularly in certain types of cancer. Even a catheter can irritate the bladder lining and produce reddening of the urine but occasionally what looks like blood staining is actually something as innocuous as a dye in some of the common laxatives or even the result of eating beetroot! There are many treatments to reduce or stop it, ranging from special tablets to solutions instilled into the bladder via a catheter.

A person with leukaemia or cancer affecting the bone marrow is more likely to bleed or have skin bruises than other people and carers have to be prepared to handle such patients most carefully. In those patients the professional attendants will give as few injections as possible because there can be modest bleeding at injection sites.

Frightening as all this must sound, it is worth emphasizing that serious haemorrhage, and certainly one so bad as to be life-threatening, is rare but is *always* something to be reported to and discussed with doctor.

Paralysis

Nearly everyone becomes weak and frail towards the end of life but what we mean by paralysis is a total loss of power or strength, usually in one or more limbs. It is not just weakness; no matter how hard a person tries the limb will not move. The cause is something wrong either in the opposite side of the brain (as can happen after a stroke) or in the spinal cord from which nerves go into the arms and legs instructing them to move as desired.

When a person has a stroke (more correctly termed a cerebro-vascular accident) it is usually one side of the body, both arm and leg, which is involved. Recovery very often takes place but it can take months and often there is residual weakness. When the right arm and leg are paralysed by a stroke, this is usually associated with such speech problems as slurring of speech or difficulty in finding the right words. There are three causes of such strokes, all much commoner in older people particularly if they have high blood pressure, hardening of arteries, or diseases of the heart valves from which mini-clots (emboli) are released. They are haemorrhage which is often so severe as to be fatal, a local blood clot in the brain (thrombosis), or a migratory embolus from the heart. People already relatively immobile or frail with other conditions are more prone to strokes than healthy, active people.

Those with cancer which has spread to the spine are liable to lose the power, not of an arm and a leg as in the stroke, but of both legs (paraplegia). The cause may be a collapse of the spine because bones are weakened with cancer deposits or cancer encroaching into the confined space where the spinal cord carries nerves from the brain to the periphery of the body. The development of paralysis is a serious and incapacitating complication for someone already very ill.

The first sign of danger, even before the paraplegia itself, is inability completely to empty the bladder, often followed within a day or so by tingling sensations then numbness in the feet and calves. Often within a day the legs will not move but even this may go unnoticed or unreported because it is all put down to the general weakness of the person. Sadly, delay can make the difference between treatment with some possibility of saving power and no treatment and no hope of correcting the condition at all. It depends on whether the diagnosis is made within 24 hours. If it is, the doctor may arrange emergency surgery or radiotherapy to relieve pressure on the spinal cord or, if neither is readily available, will himself give injections of special steroids.

Once the paraplegia has occurred, and been found to be irreversible, the question is whether or not adequate care can be given at home or permanent admission to hospital or hospice be arranged. What are needed are sufficient attendants to move and lift the patient, special mattresses and pads to prevent bedsores, the insertion of a permanent bladder catheter, regular and frequent nursing care to the skin with turning every four hours to avoid sores, and a plan to ensure bowels are kept emptied when the patient has no expulsive power of his or her own. A tall order and yet remarkably, to their great credit, many families rise to this challenge and do in fact successfully nurse such a person at home. It may mean much concerted, coordinated care for three months or so in the case of cancer patients.

Fits

A fit (usually called a convulsion) is caused by a part of the brain sending off an abnormal electrical impulse. The result may be something as mild or relatively harmless as a momentary lapse in concentration and consciousness, a jerking movement in a hand, arm, or leg, or the so-called 'grand mal' fit. This has the full blown picture of loss of consciousness, jerking and shaking of the whole body, then stiffness and finally flopping, flaccid limbs until awareness and alertness return some hours later. The patient may wet himself or herself when unconscious, occasionally has some sort of warning of the fit coming on, but afterwards has little if any recollection of what has happened. No-one will deny that fits are alarming, particularly for those who have never witnessed them before. Contrary to what might be expected, they are, in themselves, not usually life-threatening, though a person can asphyxiate if their tongue is allowed to slump back and block the airway of the throat. What matters is the underlying cause, the medical treatment and prevention, and the first-aid given by the attendants.

Of course a person long known to be an epileptic may continue to have fits when they are seriously ill with other conditions such as cancer, heart disease, or kidney disease. Others who have never had a fit in their lives may develop them as a result of cancer seedlings (metastases) going to the brain or perhaps a cancer originating in the brain. Other causes include head injury or brain surgery, kidney failure (uraemia), infection of the brain and its coverings (encephalomeningitis), biochemical upset in the blood for a variety of reasons, and different types and sites of cerebrovascular accidents.

Although it is easy to reduce the frequency of fits, or even to stop them completely with a variety of tablets, doctors will always thoroughly investigate a patient suffering them. They do so for good reasons. If the cause is a solitary cancer in the brain, particularly if it seems to have originated there rather than being a seedling from elsewhere, there is sometimes a possibility of removing it surgically or, if there is one or several seedlings, reducing their size by radiotherapy (though this will lead to hair loss on the scalp). This is why investigations are pursued even in quite ill people if the results might enable useful palliative treatment to be given but, when fits first occur in someone already ill with far-advanced cancer, the doctor may prescribe something to reduce fits and not proceed further with investigations which might only exhaust the patient. What can relatives do at home? Easy to say, but difficult at the time, 'Do not panic!'. Leave the patient as and where he is except to put his head on one side to help his breathing and loosen any collar or tight clothing, if possible putting a soft pillow or cushion under his head. Do *not* attempt to restrain his arms or legs and do *not* shout or shake him. Sit beside him until he is peaceful again. Only if the patient has another fit before he has come round from the previous one need a doctor be called as an emergency. To help in case that happens, the doctor may leave a small plastic container of a drug in the house to be squeezed into the patient's rectum via a nozzle. There it is absorbed and within ten minutes can prevent further attacks. *Fits need not be a reason for giving up on home care.*

Breathlessness

As we all know, there are many causes of shortage of breath or difficulty in breathing, but whatever the cause the experience is frightening both for the patient and for the onlookers who feel so helpless. The problem is particularly associated with heart disease and cancer, in both of which it can occur suddenly with little warning. Thousands of others endure breathing difficulties on and off for years when they have bronchitis, emphysema, asthma, or what we now more accurately refer to as 'chronic obstructive pulmonary disease' (COPD). Because they have had it for so long they have become accustomed to it, disabling as it can be.

Here we are looking at crises—sudden attacks of breathlessness when it feels and looks as though there will not be another breath. Whatever the cause, and it is usually easy for a doctor to define it, three things must be done at once.

The first is that the person should sit up or be propped up in bed or a chair before anything else—on no account should they lie flat when they have slipped down off their pillows and cushions. Secondly, the doctor should be called if the patient is not much better in 30 minutes. Thirdly, everyone in attendance must make an effort to remain as calm as possible. Nothing is more liable to worsen the panic of the patient than agitation and panic in the family! Open a window, switch on a fan, prop the person up, sit down beside him, hold his hand, and keep quiet! The doctor can do much both for the attacks themselves and also to reduce their frequency and the inevitable panic they produce.

Confusion and suspicion

Confusion is a common feature of many serious illnesses but is seldom expected by carers who immediately blame the medication or assume it is 'mental' illness. The causes are legion and the doctor will always try to define them. Often he will not find a cause but will still be able to advise. The carers must not fall into the trap of correcting the patient's mistakes and misunderstandings or rebuking him for what appears to be stupidity or stubbornness. If the patient cannot remember, or make sense of what is happening around him, it will make it worse if someone shouts at him or addresses him like a recalcitrant child!

Forgetting the time of day, or what they had to eat, is bad enough but when the patient has paranoid symptoms it is profoundly upsetting for everyone. They suspect their carers of poisoning them, of neglecting them, or of trying to harm them. A husband of 40 years or more may, out of the blue, accuse a wife of infidelity . . . the permutations are endless and very distressing. The doctor must be informed at once. There are drugs he can prescribe and others he may want to stop. Immediate sedation is usually necessary, and should a family feel they cannot cope, hospital admission is the appropriate emergency action. Fortunately, much can be done and return home is often feasible. Interestingly such patients often have no recall of these episodes and clearly no-one would wish to remind them.

10

The unexpected

A recurring theme in this book is that good care can usually be planned. The experienced doctor or nurse, whether in hospital or home, has a fair idea of what to expect, what the patient may suffer. They ought to know how to prepare and plan for such eventualities. Sad or even devastated as the carers will be, it is unquestionably easier for them to cope when they, too, know both what is happening and what may happen.

However, unexpected things do happen. They may not worry the doctor or nurses but can startle and upset relatives and friends. No blame should be attached to the professionals for not listing *everything* which might occur because to do so could be even more frightening. Would people cope better if they knew the patient might become confused or suspicious or if they were warned of the increased likelihood of bleeding, fits, or strokes?

This chapter looks at some of the unexpected things which might happen but which do not affect all patients, relatively common as many of them are.

Temporary remission

By a remission is meant a short-lived betterment—not a cure but rather a temporary slowing down of the disease progression for weeks or even months, during which time the patient often both feels and looks better. There are various objective tests which a doctor can do which will be able to show that while the disease has not been cured, the process is temporarily less active and life-threatening.

Most chemotherapy is designed to achieve such remissions and so in patients receiving it we can hardly regard them as unexpected. More surprising, and equally welcome, are those which can follow skilled palliation of symptoms—reduction in pain, easier breathing, better appetite, fewer disturbed nights, and so forth. Staff working in palliative care units (hospices) are very familiar with this phenomenon. Patients are admitted suffering so much with their advanced disease that everyone, including the patients, expects them to die. Within days they are more comfortable, and within a week or two may be returning home. Sometimes they not only look better but their tests (X-rays, blood tests, scans, and so on) may all confirm this temporary slowing down of the disease.

There are many other patients in whom no explanation can be found for the remission. Some will hypothesize that it was their own determination or willpower which achieved it. Others will credit the return home, the visits of loved ones, the benefits of complementary medicine, the answer to prayer . . . we simply do not know the true reason.

It might be thought that every disease remission or subjective improvement would be a welcome event but that can be far from the case. The patient will still die, though later than was originally thought. There may still be more weeks or months of weakness, suffering, new crises, and difficult decisions, all taxing the carers as well as the patient. Carers, without always realizing they are doing so, programme themselves for a certain period of stressful caring, perhaps the time they thought a doctor 'gave' the patient. When it stretches out far beyond that time, they begin to doubt if and how they can possibly cope. I have encountered hundreds of caring relatives who have tearfully confessed to feeling upset or even disappointed when the prognosis was reassessed to be much longer than expected. 'Of course I am happy for him, but how on earth will *I* cope—I am exhausted now and the prospect of going through it all again and for him still to have to die at the end of it . . .'

The thing to do is to discuss it with the family doctor, honestly ventilating every feeling and fear without embarrassment or guilt. There is, after all, nothing to be ashamed of. You are still as caring and loving as before but anxious in case you run out of energy or cannot take time off work. Your doctor (or often the patient's doctor) will not be able to give you stamina but may well be able to arrange a holiday or respite stay somewhere for the patient, extra help in the home, perhaps someone to sit in with the patient while you go out, or some other form of help. Even talking about how you feel can help.

Personality changes

It is well recognized that cancer seedlings (metastases) going to the brain can produce personality changes, even when the patient may look remarkably well in all other respects. Other causes can be minor strokes, chemical changes in the blood, or simply the steady advance of any debilitating illness. Whatever the cause which the doctor finds, it can be a most upsetting experience for the carers. Their feelings will alternate between embarrassed surprise and deep hurt that he or she now behaves like a different person, almost a stranger. As so many have said to me, 'I find it difficult to look after him now because I don't feel he is my husband any more. He is a different person.'

The changes take many forms. Some patients become disinhibited, so casual and careless with their dress or behaviour that it looks as though they are exposing themselves and trying to shock. Perhaps they use vulgar or obscene language, quite foreign to them. Others develop dirty habits when they are eating or are rude to visitors.

Not all changes are for the worse. Many people become quieter and more gentle, more thoughtful and considerate. Others show appreciation when previously they were regarded as selfish people. One woman asked me what pill her husband was getting which had so changed him for the better—'My life with him has been a misery for thirty years and now he is as nice as when we were courting'.

The situation simply has to be accepted for better or worse. No-one is to blame and nothing will reverse the process, though patients with brain tumours may improve temporarily. Rebuking the patient, correcting them when confused, or rejecting them does not help anyone. It is important to realize that these are not true mental changes, not the sort of conditions which might normally have necessitated their admission to a psychiatric hospital. Once again, it can help to talk about it with doctor or nurse. They are certain to understand and will help the carer to cope.

Incontinence

To many carers it seems the last straw when the patient loses control of bladder or bowel, or indeed of both as sometimes happens. Dignity is lost, the hope of continuing to look after him or her at home begins to look remote, and the final helplessness of the situation is all too evident.

This is discussed in Chapters 9 and 15 but many people at home do manage to care for someone with urinary incontinence when they have been catheterized or supplied with special pads. Such a complication need not mean the end of home caring. So often can faecal incontinence be helped that a short admission to hospital or hospice is usually recommended and when he returns home the services of a special laundry can be used to deal with soiled linen. What matters is that each case be reported and investigated and no blame attached to the patient who will inevitably be embarrassed but blameless.

Convulsions

These are discussed more fully on p. 48.

Haemorrhage

Catastrophic bleeding is rare, contrary to what most people think. Perhaps our image of death is too conditioned by TV drama, newsreels, and opera! Nevertheless, rare things do happen and can be alarming, especially if the patient is at home. In hospital there are always staff equipped to cope with it. At home it should be reported immediately to the doctor. When the cause is known the doctor may be able to stop or reduce it in various ways, yet manage to keep the person at home. When it might get worse, he will nearly always be admitted to hospital. If the patient is near to death, and moving to hospital would be not only unhelpful but the final indignity, the doctor will so tranquillize the patient that bleeding lessens and he is totally unaware of it.

However, what needs to be remembered is that serious external bleeding, that is bleeding obvious to the carer, is so unusual that the possibility of it need never be a reason for not caring for someone at home.

Smell

A few cancers, both on the surface and deep in the body, can produce foul odours which permeate a room or ward. They can *always* be reduced or even eradicated with various preparations and equipment prescribable by the doctor. Do not be embarrassed about mentioning this problem.

Sleep disturbance

Many ill people find sleep elusive. Sometimes the cause is pain or discomfort because of pressure sores, a badly made bed, or even crumbs in bed; other causes are a full bladder, fears, or bad dreams. All can be helped by the doctor and often the prescription of a gentle sleeping pill will help.

More distressing for the family when they are looking after him at home is when the patient seems to turn night into day and vice versa. He sleeps all day and is awake and agitated all night—something occasionally seen in patients with brain tumours. A special sleeping tablet is needed even though it may make him sleepy day and night for the first few days of use. This problem must be reported to the doctor because otherwise the carers will soon be exhausted.

Unexpected decline

We have already spoken of unexpected remission or improvements; now I will discuss the exact opposite. Just when everyone expects him to live for weeks or months, for no apparent or discernible reason the patient goes downhill and he dies within days. This is a particularly common event in hospital or hospice, even when the admission has only been seen as a 'respite' to give the family a breather. Understandably the carers assume he has been given some fatal injection or overdose but this is not the case.

There are two explanations. The first is that the patient was actually more ill than anyone appreciated. The family have coped as long as possible, in fact until they were utterly exhausted, and should really have agreed to a respite admission long before. Though this admission was spoken of as a 'respite', if they had looked more carefully they would have agreed that he or she was really very, very ill and much nearer the end than anyone liked to admit.

The other explanation is one I have often seen. The patient has kept up appearances *for the sake of the family* and on getting into the hospice *consciously* lets himself go. How often I have welcomed a patient who, within an hour of settling in, has said, 'It's good to be here—I've tried so hard to keep going at home for the sake of the family but I feel I can go now and they won't have the worry or the work'. Each hour they get frailer and frailer yet remain serene and dignified, totally in control. They know they are dying and are quite resigned to it. It is not defeat or throwing in the sponge—rather the opposite. The patient is ready to die but the family is deeply shocked and bewildered and inevitably blames the doctors.

This is yet another reminder of something well recognized by doctors and nurses but mystifying to most carers. People seem to know when they are dying and behave as if it was a 'normal' event. It is those who lose them and want to keep them longer who persist in thinking of death as a disaster, a medical failure for which one day there may be a cure.

11

Unmentionable feelings

Nearly all of us find caring for the terminally ill not only difficult and demanding but also distressing. Certainly the professionals do and are able to say so. How much more difficult it must be for relatives, which is why I have written this book. This chapter will look briefly at feelings they may experience but find it well nigh impossible to speak of, even to those nearest to them. They feel that to do so might make them appear callous or selfish, uncaring, or downright inhuman. You may at first be shocked if you have never yet experienced such feelings. Describing them does not imply that everyone experiences them but merely that many will do so without ever appreciating that they are by no means unique.

Anger

It is difficult to believe that anyone can lose someone they love and not at some time feel angry. Fortunately, most people can express their anger and then get over it and continue with their caring, recognizing that continuing anger can be destructive and divisive, of little help to anyone. Expressing it can be helpful, provided it is not continually directed at the one target, particularly if that person or organization did not deserve such an onslaught.

Often the anger takes the form of blame. People tend to direct their anger at all and sundry in a random manner. At one time it may be against the doctor or hospital, at others against society and the local community for its poor facilities, and, yet again, against the church, against God, and against what are seen as God's inadequacies and faults. Nearly everybody at some time or another blames themselves for 'failing' the patient.

Doctors are often blamed for failures in communication—or the messenger is blamed for the message he carries. That is to say, the doctor is blamed because of the cancer or his diagnosis of it, or how he explained it, as if he was to blame for the patient having cancer. It sometimes happens that an otherwise caring relative will even be angry with the patient for having cancer as though he set out to acquire it. The distraught wife exclaims, 'Why did he have to go and get cancer just when we need him!'.

We are all familiar with how easily we blame those who are nearest and dearest to us, those who are doing most for us. Many of us in health are charming to the world but far from pleasant to live with. Some dying people are like this—they give their families absolute hell but are sweetness and light to the professionals!

No-one would pretend that coping with anger is easy, either for the person with it or for those around who may be the targets. However, it must be expressed, the reasons explored, and then be left behind because its presence inevitably hinders and impedes effective caring. In the end it may so damage or destroy relationships that life for the survivors will not be worth living. Likewise, anger against doctors and nurses must not be allowed to continue no matter how justified the blame. If it does, caring is made almost impossible when any action, every prescription, everything said is subject to criticism and blame.

Resentment

Related to anger, but less dramatically exhibited, is resentment. It may be resentment of the extra work or caring involved, of the sacrifices having to be made, or the attention the patient and others are getting.

Understandable is the quiet resentment sometimes felt by the unmarried daughter who has so uncomplainingly and so faithfully looked after elderly parents when other members of the family could easily have done more. At the end, when she is exhausted, they come on the scene and 'take over', telling her and the doctors what should be done, without any acknowledgement of her devoted care over years and years. As if to add insult to injury, they are often greeted by the dying patient like the returning prodigal and the will changed in their favour at the last minute!

Some resent not being given the confidential information about the patient they feel they have a right to but which the patient has not given permission to divulge. Others feel too much is expected of them by the patient who has seldom been considerate to them: 'I'm expected to give up my job and do everything for him after 30 years during which he never once thought of me and the children'.

The end of someone's life is seldom the time to tell the patient of the bitter resentment you feel but it can be talked over with a doctor or nurse, either at home or in the hospital or hospice.

Fears about the future

A feature of everyone's grief before a death is personal fear about their future. What will the future hold? How will they cope alone or without their lifetime's partner? Will life be worth living? Where will they live, what money will they have, how will they make decisions, who can they turn to?

Only in a deeply loving relationship can these fears be talked about between patient and partner. In most it will sound very selfish, an intrusion into someone else's suffering and loneliness.

The longer a couple have been together, the more their lives and interests have become so interwoven as to appear inseparable. They have done most things together, have shared friends, and cannot imagine living a 'solo' life. When an elderly person loses their life partner, they feel that all they have to look forward to is increasing frailty and dependency, alone and sometimes friendless. Few of us are good at coping with major changes in our lives but the death of a partner often heralds not only aloneness and loneliness but a move to another house or old people's home, new neighbours, and a different community. How natural it is to fear the future.

For some reason, caring relatives feel that they should not speak about their own future even though it occupies so much of their time. There is nothing to be ashamed of in feeling this way. Talk it over with the family doctor, clergyman, or close friend. Do not put on a brave face but let others know your feelings which in no way can or should be interpreted as uncaring towards the dying loved one. How true is the saying that for the patient death is the end while for the relatives it is but the beginning of a radically new life. We all know this so why not speak about it?

Helplessness

Faced with anything new in our lives, we all wonder how we shall cope. Being asked to care for someone dying, to do simple nursing, respond to his questions and changing moods when it is all new, is all daunting.

It is important to realize that *everyone* feels helpless and ill-equipped. They see the skill of nurses in hospitals and the quiet efficiency of doctors and cannot imagine how they could do the same if the patient ever came home. How would they feed him, turn him in bed, take him to the toilet, and do the most personal things for him? Many have never been near death or seen a body. They feel helpless and soon find themselves asking that the patient be kept in hospital where there are experts trained for this work. It is not the workload which people fear but failing the patient. Is it really possible to 'fail' when care is based on love?

Once again, the answer is to admit to this feeling with the doctors and nurses and to let them rehearse with you all that will be involved. Ask what outside help can be provided, and how your needs (as distinct from the patient's) can be addressed. As this book sets out to show—we can cope when we must and usually do so eminently well.

Losing faith

Though many people today would deny having a faith, many nevertheless have something resembling faith, some ill-defined concept of a God sufficient to make them pray for a reprieve for the loved one. Others come to this crisis in life with a deep faith which they confidently expect will see them through.

For a great variety of reasons, this faith may temporarily be lost, but how can they speak of it when so many see them as believers? The minister or priest and friends from church seem to assume that their faith is unshakeable, yet each day God seems further and further away from them. They pray for a cure and God seems deaf to their pleading. They pray for strength yet feel more exhausted each day. They search for meaning in what is happening and find only helplessness and pain. Where is their faith? Did they ever have a faith? Have they been lifelong hypocrites, they begin to ask themselves.

It should be a comfort to know that this happens to most if not all people. It is the exception for people to regain or reawaken faith at this time, but it has not been lost for ever. As the months and years pass, they, like thousands of others, will find that faith returns in a more mature form, tested, tried, and strengthened by what they have gone through. Trite as it may sound, it does not mean that God is not there nor unwilling to help or hear because He doesn't respond as we would wish. The sun is still there even when temporarily hidden by a dark cloud. Perhaps our problem lies in having had an immature faith which always seemed adequate for the mountaintops of life but turns out to be less than adequate for the dark, shadowy valleys like this one.

Secret dreams for the future

When so many marriages now end in divorce, and so many others seem to be just shams devoid of love and respect, it should not surprise us if death may promise a welcome, but secret, release from such bondage, and offer the possibility of long-searched-for happiness. Who would dare to speak of such thoughts, even though many have them?

So often in my work with the dying and their partners, I have been told of such marriages. One woman described how she and her husand of many years had decided to separate and start divorce proceedings. On the very day they were to see a lawyer, her husband was told he had a lung cancer liable to kill him within a year or so. 'How could I leave him then?' she asked. She stayed on to care for him and this she did respectfully and efficiently but, as she told me, 'I could not forget the past nor stop thinking that when he had gone I might find the happiness that had eluded me for years.'

Some secretly hope to find a new partner, others to start a new life with a lover, or resume a long-lost sex life, while others again look forward to returning to a career or undertaking new training.

There are other implications for these jumbled emotions. Some find they cannot cry yet society expects them to. Others find hospital visiting a hypocritical duty. Others cannot speak and behave in the sad, heartbroken way people seem to expect. Their only help lies in sharing these secret thoughts with the family doctor or hospice staff who, not surprisingly, have heard it all before on many occasions. Perhaps they need someone to give them permission not to visit each day or not to stay too long. Perhaps they need to tell a nurse how they enjoyed a day off or broach with the doctor the subject of a holiday. These carers are normal!

Finally, there are those who look at the patient and see themselves in a few years' time. They see someone dying of cancer and wonder if that lies in store for them. They are taking heart pills themselves and have to sit by the bedside of someone dying of a heart disease. Is that what lies in store? They have HIV and sit gazing at the wasting death of a partner with AIDS. How many years will it be before they are in that same position?

We cannot avoid these issues nor escape these emotions. We are human and frail, often frightened and lonely but, for some reason, are usually much helped by ventilating these feelings. We most certainly do not always require 'counselling' but rather skilled empathetic listening. There are many able to offer this—doctors, clergy, Macmillan nurses, and social workers. Denying such disturbing emotions never helps. When people discover that others have felt the same—whatever it is—they find it easier to face another day of caring and losing.

12

Aids and equipment for home care

It is quite possible to care for someone at home without any or all of the sophisticated equipment to be found in hospitals. What matters is the willingness to care. Having said that, it has to be admitted that the availability of some aids and equipment can make it easier for all concerned.

This chapter looks at such aids and how they can help. Details of where and how they may be obtained are a matter for the family doctor and community nurses to advise. Hopefully even learning that such aids exist may encourage more to offer care at home.

Mattresses

Anyone lying or sitting in one position for any length of time is liable to develop a pressure sore. It results from diminished blood circulating in the tissues compressed between the bed (or chair) on the one hand and their bones on the other. No other 'injury' need be involved. One day the skin of the bottom, heels, or elbows may look healthy, the next day very red and tender, and by the following day there is a hairline crack in the skin. In no time it will have extended and within a week or so can be palm-sized and down to the bone. Healing, if it occurs, can take months. Never was it more true to say—prevention is preferable to cure. How, then, do we prevent pressure sores?

The simple answer, but far from simple to put into practice, is not to allow a person to sit or lie for any length of time in one position. This is even more important than daily or more frequent skilled rubbing of the skin with soap and water.

The relatively well patient can be encouraged to change position, to get up and walk for a minute or two, and to move to a different chair. The frailer ones will need special pads for chairs and, particularly when bedbound, one of a range of mattresses for the bed. Some are synthetic sheepskin put on top of the bed drawsheet so that the person lies directly on top of it. They can be purchased or borrowed from community stores. More sophisticated ones, 'bubble' or 'ripple' mattresses are designed to keep shifting the site of pressure. A popular one in Britain is a Spenco with an excellent record of preventing sores and with understandably good patient acceptance. A few community organizations lend them but they can also be purchased, in which case the doctor will issue a certificate for VAT exemption.

Specially shaped pads with Velcro fastenings are available in synthetic sheepskin for elbows, knees, and ankles. As with all such aids, they should be used as preventions and not when sores have already developed.

Delta pillows

Sometimes called 'V' pillows, these are V-shaped and tightly filled to support the patient who needs to be propped up in bed. Ordinary pillows are still needed but not as many as would be the case without the delta pillows, which give such good support. They are invaluable for the person who keeps slipping down in bed.

Wheelchairs

There are many different types and models of wheelchair to cater for those with different disabilities and needs. Whether or not one would be useful will be decided by the doctor, nurse, or occupational therapist; the patient is then measured (and sometimes weighed) before the local stores will supply and deliver on loan. It is important never to go out and buy one until fully advised by the experts. Always state whether you hope to take the patient out in a car so that an appropriate model can be supplied to fit in the car boot.

Walking aids

Whether a simple walking stick or a sophisticated zimmer complete with wheels and brakes, all such aids should only be borrowed or bought after careful assessment by a physiotherapist.

Nylon ladder and 'monkey poles'

Both of these are designed to help a patient pull himself or herself up in bed, often a difficult and tiring thing to do, but it can help the patient to feel less dependent on others than he would otherwise be.

The nylon ladder is a simple rope ladder, one end of which is fastened to the foot end of the bedstead. The other end is held by the patient who can then pull himself or herself up the rungs of the ladder. The so-called 'monkey pole' is familiar to all who have visited hospitals. It is a handle which hangs over the heads of the patient and, provided both his arms can move and hold it, he can pull himself up. Both can be borrowed from community stores and used with ordinary domestic beds.

Toilet and bathroom appliances

After a thorough home assessment, a community occupational therapist will sometimes recommend and arrange to supply aids to raise the toilet seat, handrails at the side of the toilet to facilitate getting up and down, and whatever else is needed for the bathroom —shower fitting, seat in the bath, handrail to aid bathing, and much else. The occupational therapist is usually invited in by the doctor or one of his colleagues.

TV and radio

It is not a luxury to have a remote control for a bedroom TV set. It gives patients control over their environment and saves them feeling a burden or nuisance, always having to ask others to help them. For quite a small cost, a TV can be fitted with a device and the patient supplied with an appropriate headset which will allow him, or rather the others in the house, to view TV without sound being heard by others. Such a modification to the TV set in the lounge can obviously save disturbance to the patient in the nearby bedroom.

A useful, cheap device is an earpiece which can be fitted into the headphone socket of a portable radio or CD player, and then put under the pillow. The frail patient can thus lie in bed listening to music of his choice, unheard by anyone else, yet without having to wear headphones. Many people find music more naturally relaxing than any sedative or tranquillizer prescribed by their doctor. When friends ask what present the person would like, it is good to be able to ask for a tape of their favourite music rather than flowers or chocolates!

Talking books

These have long been a boon for the blind. There is now a large range of audio tapes and CDs of readings, praise, and poetry available for anyone with a player.

Here it is worth remembering the benefit of a telephone near the home-based patient. There are many different models and adaptations to choose from, with amplifiers for the hard-of-hearing, large buttons for those with hand disabilities, special appliances to hold the receiver, and so on. Local showrooms are always happy to advise and a grant to have a telephone installed can often be obtained from one of the major cancer charities.

Speaking and communication aids

Patients who have for some reason lost the power of speech have usually been thoroughly advised about speaking aids by the speech therapist.

There are others who have lost not only speech but also other powers. For them there are many highly sophisticated appliances which can simulate speech or print it out. Some are controlled by hand, others by breath, eyelid movement, or other means. Here it is sufficient to say that they exist and when someone has an illness likely to affect communication the sooner these devices are discussed and provided the better. Even the most sophisticated can be used and fitted in small domestic bedrooms. Never forget the benefit of a simple battery-powered room-to-room intercom. Easy to install, it frees the carer to work in another room and still be in touch with the patient.

Fireproof covers

The danger of fire is ever present when a cigarette smoker insists on smoking when he is frail and liable to drop the lighted cigarette on to the carpet or the bedcovers. A fireproof cover shaped like an apron is available, equally useful whether he is sitting in a chair or in bed.

Crockery, cutlery and liquidizers

A range of special plates, cups, knives, forks, and so on is available for different disabilities. The advice of a doctor or occupational therapist is not essential before buying but is certainly recommended.

Oxygen

Any family doctor can prescribe oxygen and the equipment required to administer it for use in the home. Suppliers deliver it to the house. Sophisticated machines known as condensers which do away with the need for oxygen cylinders are expensive and can only be prescribed for a patient at home by a consultant in chest medicine. They save having to store bulky cylinders in the home but can be noisy.

13 ....

Ethical issues

The title of this chapter, 'Ethical issues', might frighten the reader who feels he has enough to worry about without having to think about ethics. It is only when one begins to think about confidentiality, the rights and wrongs of withdrawing treatment, or perhaps initiating treatment against the wishes of a patient that the relevance and importance of ethics become apparent. If for some reason the carers allow themselves to think about euthanasia, assisted suicide, or living wills, then the need for this brief chapter becomes even clearer. Ethical issues are complex but unavoidable though only the sketchiest of descriptions can be given here.

Confidentiality

This ethical principle has inevitably been alluded to elsewhere in the book. It states that no information about a patient may be passed on to anyone by a professional caring for him without the patient's informed consent. It is generally acknowledged that the professionals themselves may exchange relevant information between each other if it is in the patient's interest. This all sounds very reasonable. It is the patient's life, illness, treatment, and destiny—not someone else's—so what right has anyone else to confidential information about them? The same principle should apply to the passing on of personal details between any of the carers and those who do not need such information.

In reality the principle is breached every day. A spouse will expect to be told everything about his or her partner without having to seek permission. They often go further than that. Particularly when the illness is so serious as to be life-threatening, they expect to be told everything whether or not the patient agrees or has even been told, and then usually put an embargo on the patient himself or herself being told! Presumably this is intended to protect the patient and should be seen as a well intentioned act of love. Whatever the intention, the end result of such a conspiracy is usually hurt, suspicion, and fear.

One would like to think that in any good relationship the patient's partner would always have been given implicit or explicit permission to share information. What of other family members? Just because someone is seriously ill in hospital, possibly dying, has any relative or friend the right to telephone in for confidential information? Many do, and sound deeply hurt or aggrieved when they are not given it, but they have no *right* to it—unless the patient has expressly consented. Sometimes confidential details are divulged unintentionally by church visitors, domestic helpers, and others who mean no harm but may still hurt others by thoughtless conversation.

Readers might be surprised at this until it happens in their own lives. Their relative is upstairs in bed. When the family doctor comes downstairs, he is whisperingly invited into the lounge where many of the family are gathered, all waiting to know things which, so they have convinced themselves, the patient neither knows nor needs ever to be told! Even when people would not dream of asking the doctor to breach confidence, they will never hesitate to question the nurse and put her into a very embarrassing position. What must it feel like to hear the family whispering secretively downstairs and know that it is all about you but nothing is being said to your face? Good care is founded on mutual trust, not on suspicion and doubt.

Do not put pressure on doctors and nurses to divulge confidential information but base all your caring on love and mutual respect.

Advanced directives

In some countries the term used is living will but both mean the same, a recorded and witnessed wish of a sane person that when the time comes that he or she is so ill as to be incurable with no quality of life, nothing should be done to resuscitate or artificially maintain their life. Naturally the wording varies from document to document and country to country but basically it is designed to respect patient autonomy. In most countries it is not legally binding but nevertheless has value in that it gives insight into the thoughts and wishes of the patient before they reached this late stage of a mortal illness.

There are good arguments for and against legislation to make such advanced directives legally binding, but what cannot be denied is that these directives or wills reflect a deep horror that some life-support systems or energetic life-saving or resuscitation measures might *not* lead to a life which is both prolonged and still dignified.

Suffice it to say that in Britain, most doctors would genuinely appreciate knowing what their patient has written and would usually endeavour to respect their wishes but could not be forced by law to abide by it.

Suicide and assisted suicide

Every suicide is a tragedy. Whatever prompted a person to take their own life it must challenge everyone concerned to ask what went wrong and what more each could have done. When someone has a mortal illness of the type we are concerned with in this book, we must ask if they had unrelieved pain, unanswered questions, unventilated fears, or whether they were lonely, feeling a burden on others, guilty, or utterly hopeless.

It surprises many to hear how few terminally ill patients even with the strength and means to do so actually attempt suicide. What is not surprising to those of us who look after such people is how, rather than talking of suicide, they make such efforts to live lives to the full—*when they are freed from pain and suffering.* How true is the old adage 'pleas for euthanasia or suicide attempts are actually cries for better care'.

We shall look briefly at the question of euthanasia later, but closely related to it is the issue of so-called 'assisted suicide'. By this is meant the means being deliberately provided to a patient for them to end their own life. Clearly it differs from active euthanasia in that the person can choose to commit suicide or leave the drug, or whatever else was provided, for a subsequent attempt. So-called control is in their hands.

In Britain, suicide is not a crime, while assisting a person to commit suicide is a crime. How sad that people should want doctors to serve a dual role—one day caring and respecting life, the next leaving a sufficient supply of drugs to end that life. If we all provided good care, there would be no ethical case to answer.

Double effect

Almost every medicine or drug has several effects. One is the therapeutic effect for which it is prescribed, the other the adverse or side-effects which must be well understood by the prescribing doctor. Naturally he strives to utilize the useful effects and not the adverse effects, some of which could in theory be potentially dangerous.

This delicate balance worries some patients and their families. They appreciate that sedation is required but worry lest it will eventually kill the patient. For that reason they may suspect a doctor of wanting to shorten the patient's life when he prescribes a sedative or perhaps morphine, which so many people still fear will kill the patient as it controls his pain. Doctors do not deny that drugs may have several effects but what matters is their *intention* when prescribing them. Provided the intention was merely to sedate or relieve pain, they cannot be accused or found guilty of utilizing the less desirable actions of the drugs.

This issue is of much more than academic or philosophical importance. Very many relatives with but scanty knowledge and understanding of medicines and their uses, insist that they are not prescribed for their loved ones until the very end of life is near in case they shorten that life. In other words, these relatives deprive the patient of essential drugs to ease pain, relieve terrifying breathlessness, or in other ways bring comfort to the dying. This is understandable but inevitably has the opposite effect of all they are striving for. They are condemning the patient to totally unnecessary suffering and showing a sad lack of trust in the doctors.

Withdrawing treatment

There comes a time when nothing will prolong a life with any semblance of dignity. The disease process, whether it is cancer or some other condition, is beyond control and sooner rather than later that life will end. Discontinuing active treatment at this point is not shortening life; it is not accelerating death. It is a recognition that this life is ending and all care and attention must now be palliative, that is *directed solely and exclusively at patient comfort and dignity*. The doctor who discontinues anti-cancer drugs or heart pills is not killing or committing euthanasia but professionally acknowledging that they now have no place in the care plan.

Similarly, embarking on evermore energetic treatment when it has no possibility of extending, maintaining, or even improving the quality of life is wrong. Examples might include intensive antibiotic treatment for terminal infection, energetic intravenous fluid therapy for someone with kidney failure, or blood transfusion for a person dying of widespread cancer.

So often otherwise well intentioned and caring relatives, knowing full well that their loved one is dying, demand that extraordinary means be undertaken to 'keep him going'. They ask for intravenous fluids, injected antibiotics, or stomach tubes to force-feed him, choosing not to see how each measure would add insult to injury and come to be an unnecessary barrier between them and the one they love.

Euthanasia

As many people know, the word 'euthanasia' comes from Greek words and strictly means 'a good death'. In that sense it is what we all want for ourselves and our loved ones.

That is, however, not how the word has come to be used in modern parlance. It is now taken to mean the deliberate ending of a person's life with their full and informed consent, in fact at their request, often recorded and duly witnessed long before their terminal illness, with the intention of saving them unnecessary suffering or loss of dignity and autonomy. It is worth emphasizing that such 'active, voluntary euthanasia' does *not* mean withdrawal of treatment, switching off a life-support system, keeping alive a body which was otherwise biologically dead, nor the principle of double-effect. Expressed crudely but accurately, active, voluntary euthanasia is the deliberate killing of a patient to kill the pain! It is an illegal act, no matter how well-intentioned and ostensibly compassionate. How surprising that relatives should expect doctors to be both healers and killers.

In one sense it is like the issue of suicide. People expect patients to ask for euthanasia because in their days of health and vitality they spoke as though they would want it when their time came. They cannot believe it when the same person seems to hold on to life and never again speak of suicide or euthanasia. It is like so many firmly held opinions which many of us have at some time or other: when circumstances change, so do these opinions. I have encountered many who vehemently advocated euthanasia only to say at the end of their life how wrong they had been and how they had relished the time they have now been able to spend with loved ones—*when freed from their pain and suffering.*

Again, like suicide or physician-assisted suicide, euthanasia is a topic which can seem so important and relevant to relatives (who, incidentally, would usually not perform it themselves but expect a doctor to perform it on their behalf) but is scarcely mentioned by the comfortable patient who is the central figure in this drama.

Perhaps a true story will illustrate this. I had as one of my patients a recently retired doctor. All his life he had advocated euthanasia and now he was dying of a painful form of cancer. He asked if he might speak to a group of medical students, telling them how deeply he wished that I was permitted to end his life which now had no meaning for him. At the end of his time with the students, all fascinated and moved by such a personal account of dying, I went into the patient's room carrying some blood test results in my hand. The doctor – patient stopped his tutorial and asked what his own blood results showed. On being told how weak his blood had become, he immediately dismissed the students saying that he must have a blood transfusion at once because otherwise he would die! Without another word being said, for none was needed, he saw the irony of it all and laughed. 'Dear me, how very difficult it all is when the time comes! Let's get that blood into me!'

Let us continue to recognize that any call for euthanasia is in fact a call for better care—and strive to give it.

14

Children, pets, and other friends

Children

How much should children be told of what is happening or allowed to be with a terminally ill person they love? Will they be shocked or psychologically damaged? How can one explain these things to them in language they will understand?

Only a brief overview and guidelines can be given here but Appendix B gives details of books designed to help both the children, and those responsible for them, at this time and in their later grief.

- Allow them to spend short times with the patient, whether in hospital or at home, unless there is such physical disfigurement as to be frightening. Adults are keenly aware of someone's pallor, jaundiced complexion, or wasting, but children seem able to sense or see the real person under that external shell. It is still their grandpa or mummy and they love them for that reason.

- Prepare them by explaining how tired grandpa is, how he may be sleeping or feeling sick when they go in. Explain how you will all stay for just a few minutes so as not to tire him, or why everyone has to be careful not to jump on the bed or bump his chair.

- Do not try to deny what is happening by describing it in euphemistic terms—'He's going to live with Jesus', or 'God wants him to live with Him'. This will only make them angry with God for taking away their grandpa when they want him here with them.

- Children under six or seven have little, if any, concept of death except as separation or loss. Unlike adults, they do not look forward to a future which will be one without their grandpa.

Their whole life centres around today and the continuing security they feel and need.

Children between about seven and adolescence already know something about death, though adults might think it is very different from the death of grandpa. They have lost a pet rabbit or dog and possibly seen him buried in the garden. They remember being sad and not knowing what to do, but at least they have some basic appreciation that all life comes to an and.

Older children can be misled or taken in by our well intentioned euphemisms. Life for them is black or white, and they take what we say literally. I well remember a young boy who came back to see me a few weeks after his father had died. He was very angry—'You never told me he would die! You told me he wouldn't get better! I thought you meant he would always live in the hospice here, and now he's dead!'

- If you need help in explaining what is happening, ask the doctor or nurse if they will speak to the child or help you to find someone else to help in telling him. *Never leave it to chance in the hope that the truth will dawn on the child.* They are quite likely to come to the wrong conclusion, to suspect that terrible things are being done to him, to become distrustful of doctors or fearful of hospitals. 'Just tell him you don't know how to cure his grandpa, doctor' said one parent, not realizing how confusing and frightening this would be. 'We intend to say he has gone on a long voyage and may not come back for years', said another. What a false image that would have created in a child's mind.

- Do not be afraid to show your own feelings, even your tears, in front of children, provided through it all you maintain that image of safety and security which every child needs. Children can cope with such loss of a parent or grandparent when they themselves feel safe and loved. It is when parents are themselves overwhelmed with grief so that they ignore the needs of the child beside them that the young one is at risk.

- Trite as it may sound, children like adults can grow and mature as a result of their grief and exposure to death if it is not cloaked in mystery or myth. If they come to associate dying and death not only with sadness, but with love and caring, they will more readily grow into loving, caring adults.

- Always discuss what is happening with their school teachers so that they may be alerted to unusual behaviour or poor class-work and understand the reason for it. Teachers can play a vital and deeply sensitive role at this time and appreciate being drawn into the caring circle. Often the school will have access to a child psychologist who can help.

- Unless advised otherwise by doctor or psychologist, permit older children and teenagers to attend the funeral, provided those they love and trust are always beside them, helping them to feel safe and still loved in their sadness.

Pets

Few hospitals are enlightened enough to permit the briefest of visits by someone's pet, but most hospices will welcome them once permission is sought from the sister-in-charge.

Such visits by, for example, the patient's dog must of necessity be short and the pet kept under control but they can be of inestimable value. Many is the patient who is dramatically relaxed in his chair as he strokes his dog's head lying across his knees.

Friends

Mention has already been made of visiting at home or in hospital by friends. Their visit can be mixed blessings. The thoughtful friend stays minutes, the inconsiderate an hour; the one brings news and interest, the other either asks questions all the time or exhausts the patient with interminable chatter.

Few appreciate how tiring visitors can be and even fewer have the gift of sitting quietly and making no demands on the patient.

The immediate family or next-of-kin must take control of the situation. They must respect the patient's wishes about who can visit but when he is in danger of being exhausted by so-called friends, they must decide who is to be permitted in and who not, telling those concerned and discussing it with hospital or hospice staff. The easiest solution is to say that the doctor has said that only close relatives or 'only his wife' may come. Acquaintances will probably be offended, good friends completely understanding.

On no account should any friend be given confidential information without the patient's consent and never should a friend be allowed to 'advise' about treatment or alternative care. True friends are never upset by restrictions designed only to help the patient.

15 ....

The final days

No matter how expected a death, or how prepared the family and friends are, the end always seems to come as a surprise. This chapter is designed to give a better understanding of those final days.

Bodily changes

While it is certainly true that doctors are often unable to predict exactly how long a person will live, they are usually able to say when the patient has come to the last few days. The most marked feature is the tiredness and sleepiness, often thought by the relatives to be due to sedative drugs when it is actually the effect of the illness itself. As each day passes, he is awake less and sleeps more, often very deeply. His weariness is even more marked. Every movement, every action seems a physical effort, often requiring the assistance of those near him. Not only can he not feed himself but even lifting his head off the pillow seems too much, putting out his hand to grasp that of a loved one is difficult. Conversation almost stops and only a few words are spoken, greeting someone or asking for a sip of drink. As his voice gets weaker, those nearby have difficulty in making out what he is saying and find themselves sitting closer and closer so as not to miss anything.

When we sleep we all have short spells when our breathing sounds slower or shallower than normal, but of course we ourselves are not aware of this. As death approaches, this pattern of breathing is seen during the day, as well as at night, whenever he is sleeping. It takes the form of several seconds without any breathing, followed by a small breath, and then a slightly deeper one, and finally a deep sigh followed once again by nothing for another few seconds. This interval may only be four or five seconds but, to the onlooker, can seem like a minute. As the end approaches, it gets longer—even 15 or 20 seconds—during which, understandably, the carers fear he has stopped breathing for ever. Then there are a few sighs until eventually there are no more. The doctor refers to this as Cheyne – Stokes breathing but, whatever its name, it is something no-one can be unaware of, or not at first puzzled by, as they sit by the bedside. The thing to remember is that it is normal, automatic breathing—not fighting for breath and certainly not anything the patient is aware of.

His need for nutritious food stops and even fluids are not now wanted or taken only in sips or by sucking on little plastic sponges provided by the nurses. Once again this is right, the body signifying in its own way that it no longer needs such things to survive. By this time, if not for much longer, the body has been unable to make use of most of the food proffered. Even drinks are not so important except for the comfort of his mouth and tongue which otherwise become dry and uncomfortable. If only everybody appreciated this, but there are always some who convince themselves that he is not dying because of his illness but because no-one is attempting to feed him. They demand that he be fed by tube or a 'drip' into his arm, overlooking the fact that they would be further insults to his comfort and dignity, serving only as a barrier between the patient and those who want to be near him. The carers could best serve him by wetting his lips with a cloth or handkerchief dipped in iced water or by using one of the lollipop sponges gently put between his teeth to moisten inside his mouth.

It would be comforting to know that even at this stage the patient is still lucid, if not alert, but this is not always so. There is often some muddledness, some forgetfulness about where he is, what day or time it is, or even who has been in to see him. We should not be surprised by this. It is not 'mental illness' or the disease, whatever it is, affecting the brain. It is a reflection of his unutterable weariness and exhaustion, of his dozing and sleeping so much of the day. Do we not all know what it is like to waken after a deep sleep and temporarily, momentarily, forget where we are or what day it is? If that happens to us, how much more should it happen to someone who can no longer tell night from day. There is no question that such confusion or muddledness can, and often does, upset those looking after him, whether at home or in hospital. They have sometimes left it too late to say something of great importance, such as how sorry they are for something in the past, or how much they love him. Time after time they blame the doctors for giving him 'those drugs which made him like this', when very rarely is it the medication which, as we shall see, has usually been reduced to a basic minimum by that time. Many relatives choose not to listen to what doctors tell them. They continue to believe he can recover in spite of it being so plain for all to see that he is frailer every day. At last it seems to hit them and they blame all around them because it is now too late to say what they should have said a long time before.

Finally, usually within the last day or so, the hands and feet become colder as the body tries to keep what bodily warmth it does have in its centre rather than in the limbs. The colour fades as the skin takes on an almost transparent appearance.

At last the breathing stops, just when it looked as though he was in a deep, natural sleep; the eyes seem to gaze into space, colour drains from the lips, and everyone realizes almost to their surprise after so long expecting and dreading it, that he has died.

Medication in the final days

We have already spoken of the changes in medication suggested by the doctor. Perhaps an explanation of this will help. Many drugs can now be discontinued, no matter for how long he has been taking them in the past, or how essential they were always said to be in happier days. For example, he will no longer need drugs for angina (which is related to the amount of effort or exercise a person indulges in) when he is lying quietly in bed. Though he may for years have had tablets to keep blood pressure down, he is most unlikely to have high blood pressure in these final days and, in fact, if he has cancer that will itself have brought his blood pressure down. Even the diabetic will no longer need insulin when he is not eating much; to continue with it would be positively dangerous. The time for antibiotics, antidepressants, iron tablets, vitamin pills, and many, many more has passed. What *cannot* be stopped are painkillers, anti-epileptic drugs to prevent fits, relaxant drugs, and those which serve to dry up the bubbly secretions at the back of the throat that make the noise so distressing to relatives but unknown to the patient, often called the 'death rattle'.

Because he sleeps so much, or can't always take pills or medicines as he used to but still needs to be kept free of pain, the doctor has to decide on other ways of giving him medications. Occasionally he will prescribe suppositories—little bullet-shaped jellies inserted into the anus (the 'back passage') where they dissolve and are absorbed into the bloodstream. To many this may sound unpleasant or repugnant and many carers will feel they can't do it even for someone they love. They can ask for nurses to do it. However, there are many others who have used types of suppositories for years and find this method quite acceptable and easy and are happy to do it for the patient.

However, more and more doctors are using other means of helping. One is to prescribe little adhesive patches which resemble the sticking plasters we all use for cuts and grazes. They are actually very sophisticated devices, impregnated with exact doses of the drugs needed, which can gradually be absorbed through the skin. They are known as transdermal patches.

A more commonly employed method nowadays is to give the essential drugs via a tiny needle placed just under the skin, connected by a length of fine plastic tubing to a syringe-driver or syringe-pump. This is simply a battery-operated device so compact that it can be put in a pyjama pocket. Fastened to it is a syringe loaded with a full day's supply of the drug. The contents are gently expelled from the syringe to the needle via the tube. Of course such a device is invaluable long before the final days. In fact, many people are out and about in their homes and even at work, kept comfortable with this device which saves them having to take pills and tablets and medicines at frequent intervals during the day. Though it may resemble the familiar 'drip' so often seen in hospitals, it is quite different. The former is connected to a needle inserted into a vein, enabling vital drugs to be infused straight into the patient's bloodstream. This means, therefore, that it must be set up by a doctor and be most carefully monitored by a nurse or doctor. The syringe-driver method on the other hand is safer for home care bcause it only needs a nurse to change the syringe each day and the needle can remain in the same position for several days until being changed. It is so safe and convenient that it does not have to be monitored day and night by a nurse.

Another method of giving some drugs is by tablets specially designed to be placed under the patient's tongue where they dissolve slowly and are absorbed into the bloodstream. For such a method it is clearly essential for the patient's mouth to be kept moist, something any carer can do and, until the last day or so, assisted by putting a drop of lemon juice under the patient's tongue two or three times a day.

Whether in hospital or at home, the patient may need a catheter. This is a fine rubber tube gently inserted up the urethra (the urine pipe) into the bladder. It can be a totally painless procedure when an anaesthetic jelly is used to lubricate it and ease its passage, obviously a very short distance in a woman and slightly further in a man. It has an inflatable bulb which keeps it in place in the bladder, the urine then flowing out of its own accord into a plastic bag by the bedside or attached to the patient's leg. When someone has lost control over his or her bladder emptying and is either wet with urine or unable to pass urine (retention), such a catheter is invaluable, often restoring dignity and saving much embarrassment.

When the patient dies

It is one thing to describe a person's death but almost presumptuous to try to describe what carers feel when their loved one finally dies. The peace of it can scarcely be described. This comes as a surprise to most people who expect it to be frightening or noisy. Suddenly there is silence, peace, and all at the bedside, whether in hospital or at home, are locked in their own thoughts. Many are stunned and silent.

Some find it easy to cry, others feel as though they have dried up. Some feel an urge to speak, often to express relief. Others say they feel it's an anti-climax because, in a sense, the patient seemed to have died long before—when he became unconscious, when he became confused, or when he stopped talking to them.

Wherever someone dies, there is no reason to leave the bedside if people would like to sit together and possibly hold his or her hand, kiss for the last time, or quietly say yet again how much he or she was loved and will be missed. No-one in hospital should ever feel they must leave because the ward is busy or the staff must get on with things. This time is immensely precious, almost sacred, and to be respected.

Later, wherever he is, the extra pillows are taken away, leaving only one behind his head. Arms are brought close to the side, legs gently straightened out, and a pillow then placed under the chin to prevent his mouth dropping open. If, as is often the case, any dentures have been left out in the final days in case they came loose, they are put back in place, his face wiped with a moist cloth, and his hair tidied. Few would go so far as to say that he or she looks beautiful at this time but there is usually a striking serenity, particularly when the final weeks have been troubled ones.

What happens then depends on where the death took place. In hospital the relatives or friends will be taken into another room where they can have some privacy together, hopefully over a cup of tea, while they wait for the death certificate from the ward doctor. Often a doctor or ward sister will then spend some time with them, explaining what they have to do, such as registering the death and contacting a funeral director.

It is worth noting here that many people are so numbed with grief that they feel helpless and useless, but often will not admit to this. They cannot remember names and addresses and telephone numbers. Though normally they would have no difficulty remembering it, they can't recall the name of the minister or priest, sometimes not even the addresses and telephone numbers of close family members. This is absolutely normal, something most people experience and not anything to be ashamed of.

If the patient dies at home, the first thing carers do after attending to the body as described, is to telephone the family doctor who will come as soon as possible. He will issue a death certificate and advise what to do with it. Either then, or the following morning if death occurred during the night, he or the nurse will advise on the disposal of all drugs, particularly those for which there are special legal requirements. For some reason, doctors do not always take time to explain what they have put on the certificate but relatives should never hesitate to ask for it to be explained as it is often puzzling. Many of the terms are medical ones—for example, bronchogenic carcinoma for a lung cancer, or multiple metastases meaning that cancer seedlings had spread to many parts of the body. Sometimes several terms are recorded when the relatives had understood that there was only one fatal disease. For example, when someone dies as a result of heart failure, the doctor may say on the death certificate that death was caused by cardiac failure secondary to benign hypertension, secondary to atherosclerosis.

This is merely the correct description of the arteries being narrowed, producing a rise in blood pressure which finally caused the heart to be strained and stop.

16

Funeral arrangements

Within the next day or so the death has to be registered with the Registrar of Births, Marriages, and Deaths. In fact it can be done at any time within the first week but it is advisable to do it as soon as possible. The certificate has to be taken either to the registrar of the district in which he or she died or that of the district in which the patient lived, whichever is more convenient.

The registrar will need the death certificate and the birth and (if appropriate) marriage certificates, if they can be found. In addition he will ask for the deceased's NHS card and any social security documents but these can be taken in over the next few days.

Those registering the death will be asked if they want to purchase copies of the registration (note—not the death certificate but the registration) as documentary evidence of death for the bank, lawyer, insurance company, or anyone else who may need to see it. It is wise to obtain a few copies to save having to wait for them being returned over the following weeks from those to whom they have been sent.

The next thing to do is to contact a funeral director, either by visiting his rooms or, as most people do when someone dies at home, asking him to come to the house. If no undertaker is known or has been recommended, a name can be selected from the Yellow Pages.

The undertaker will need to see either the death certificate (if the death has not yet been registered) or a copy of the registration. He will ask if the patient expressed a wish for a burial or a cremation. If the patient did not do so, he will ask the wishes of the family. If there is to be a burial, he will need details of any plot owned by the family or deceased. If a cremation, he will make all necessary arrangements to have the family doctor complete special certificates. Here we must mention something which can at first sound disconcerting. For cremation, a second certificate has to be completed by a doctor not in partnership with or related to the family doctor. This second doctor, probably a stranger to the relatives, will come to see the body and ask a few questions of those who attended and were there at the end. This does not imply suspicion of a crime—it is a legal requirement.

The undertaker will then need to know into which newspapers notices are to be placed. It is wise to insert death notices not only in the principal local paper but in others which might be read by friends who otherwise would not hear of the death in time to attend the funeral. Perhaps the patient spent years in another town where there are still many who knew him. This notice for the papers is always drafted by the undertaker, taking into account the wishes of the family, and it is the responsibility of the undertaker to take it to the newspaper office. Sadly, there have been instances when practical jokers have put death notices in papers for people still alive.

If the family have a church connection the undertaker will, unless the family have already done so, contact the priest or minister and arrange a mutually acceptable time for him to conduct the service. The undertaker arranges for him to be picked up by car and returned wherever he has to go. If there is no family minister or priest, the undertaker will arrange with another clergyman of the deceased's denomination or, as quite often happens now, arrange for a non-religious service of thanksgiving and remembrance for the one who has died. Even the organist and music are organized by the undertaker, but always the family wishes are followed.

The undertaker will enquire how many cars will be needed in addition to the one for the principal mourners, ask for some idea how many can be expected to attend so that the right size of chapel can be booked, and finally ask if a room should be booked anywhere for a meal or reception for the mourners after the funeral service.

A few words of advice here. It has recently become the norm for people to ask 'no letters, please' but experience shows that over the next few months or years such letters can be immensely helpful and comforting, reminding you just how much he was loved and respected, how many people valued him even if they failed to say so when he was alive. Others wonder what to say about flowers. On a new grave they are beautiful and a source of comfort to everyone who sees them. After a cremation, however, they are usually thrown away, their beauty lost and wasted. For that reason, many people either ask for 'family flowers only' or request that they are taken by the undertaker to the local hospice (particularly if the patient was under its care) or that donations in lieu of flowers are sent to a charity of their choice—a lovely gesture which is deeply appreciated by any charity.

Surprising as it may be, the formalities of registering the death and arranging the funeral take only an hour or two and are much less complicated and bewildering than people expect. In fact they often say they wish they had taken longer because time seems to hang for the next few days or week until the funeral itself.

This time can be usefully spent going through address books and ensuring that all who need to be told are informed of the death. This should never be left to one person. It should be so planned that one is telephoned and asked if they will be responsible for telling some of the others. In this way, within a day, everyone knows and no one person has had the distress of having to repeat the story to dozens of others.

There is one final decision to be made after a death—whether the body is to remain in the hospital or house and be taken from there to the funeral, or whether it should be taken to the undertaker's chapel of rest until the funeral. Increasingly, people seem not to want the body to remain at home, yet that was what always happened until a generation or two ago. Perhaps they now feel it macabre to have their relative's body in the house, perhaps in an open coffin in the bedroom where he died. They wonder if they will ever be able to use that room again, and whether they or the children will have nightmares if they do. On the other hand it is where he was loved and cared for when alive, where he gave so much of himself, surrounded by treasures and memories. There is no right or wrong decision. Just do not rush into having him taken away only to regret it later. Always the question must be 'What would he or she have wanted?' and do that.

Donating the body organs

If the deceased has signed an organ donor card, it is important that the organs which can be transplanted are removed by the surgeons as soon after death as possible. In the case of a death in hospital, the medical staff should be informed in advance of the death so that appropriate arrangements could be made. Unfortunately it is more difficult after a death at home unless the family doctor has made careful prior arrangements for the body to be taken to a hospital mortuary immediately. Clearly it is *always* important that the wishes of the patient to donate organs be made known to whichever doctors are caring for him or her.

Some people want to leave their body to 'medical research' and, as advised, inform their family, doctor, and lawyer of this wish. In practice, it actually means the body being donated to a department of anatomy in a medical school. What is sometimes not appreciated is that after a person's death, the body becomes the responsibility of the next-of-kin who must decide about its disposal. Most relatives and friends would follow the patient's wishes no matter how distressing it was to them, but they need not do so. If they do follow the wishes, their responsibility is to telephone the department of anatomy immediately and have them make arrangements to remove the body (incidentally, at no expense to the family). In such cases there is no immediate funeral (but always a respectful service organized by the department when they have finished with the body), but many families have a service of thanksgiving. Finally, it must be pointed out that the department may decline to accept a body if, for some reason such as cancer, its normal anatomy is so deranged by the fatal illness as to render it of little use to them. Likewise, the body of someone known to have an overwhelming infectious disease such as AIDS would not be acceptable.

What could be done to meet the wishes of someone who always said their body was to be used for medical research yet died of widespread cancer as just described? The answer is for the relatives to give consent for a post mortem, something quite likely to be requested by the doctors. By this means it is possibly not only to ascertain exactly why someone died but also to advance medical knowledge about that disease and the different ways it afflicts people.

No-one would deny that having to decide whether or not to consent to a post mortem is deeply painful and difficult, but so much can be learned from it. It is carried out with very professional respect, usually within a day or so of the death so as not to delay funeral arrangements. Always, the doctor in charge will arrange either to meet with the family or to write to them to explain what was learned at the post mortem.

Very occasionally a post mortem has to be carried out to meet legal requirements, whether or not the next-of-kin wish it. It might be because the cause of death is not known (and cannot be deduced by the doctors), or the person died without having been under medical care, or because it must be confirmed whether or not they died of an occupation-induced illness such as the chest disease mesothelioma resulting from exposure to asbestos. In the latter case the doctor has to report the death to legal authorities who decide whether or not to demand a post mortem. *Always* the doctor will explain everything before and after such an examination.

17

Bereavement

It may seem surprising, or even inappropriate, to have a chapter devoted to the months and years after the person has died. This may not strictly be 'caring for a dying relative' but it is inextricably linked with it. In fact, the better a person is enabled to care for a loved one, the less painful and paralysing may be the subsequent grief and bereavement period.

Grief is unavoidable. It's a part of loving and losing. Much as modern society might try by so many means to avoid or deny pain and loss, it cannot do so. Sadness, loss, and grief are parts of life which may be deeply painful but are some of the experiences by which we grow.

Only a few observations will be given here. The subject is a complex one, much studied and researched by health-care workers, social workers, and counsellors, and with very good reason. 'Healthy grieving', as some have described it, is normal and can be creative. Unhealthy grieving can be disabling and very difficult to help. I have given the brief outline which follows because experience shows that people are helped when they find others have experienced something like themselves, though, it must be said very strongly, we all want to feel that our experience of losing and grieving a loved one is unique. In a sense it always is unique, but similarities with the experiences of others can help us.

Perhaps it is so self-evident that it does not need to be said. Grief begins long, long before a loved one dies; in fact it probably begins when the illness is first diagnosed, when there is even the smallest hint that one day the illness might prove fatal. As the months or years pass, so the grieving comes and goes in parallel with the illness. When things are going well, it is not forgotten but it is easier to put the fearful thoughts to the back of one's mind. As it gets worse, even temporarily, the prospect of loss looms again only to recede as health improves and the threat recedes. There finally comes a time when it can be ignored no longer. Everything points to the fact that he will die and the pain of grief becomes intense, sometimes reaching what feels like its climax when the loved one dies. It is only later that the one who is left realizes that that was not the climax. Worse has come and it feels unbearable.

The grieving before a death, in the weeks or months of intensive caring, is often made more difficult to bear by not being able to share it with the one you love—the relative who is dying. Even when there's no conspiracy of silence between them, it is clearly difficult to say how painful it is. How much worse it must be when, for some reason, the carer has decided to act as though it is not a mortal illness and tries to keep up a pretence with the one they have always been open and honest with.

Doctors and psychologists regard grieving as a type of loss or a reaction to loss. We experience many losses in our lives—the loss of a pet, the loss of a job and security, the loss of athletic skills, the loss of our own health or attractiveness, and, finally, the greatest loss of all, that of a partner through death (or divorce), and eventually the loss of our own life. We each seem to have our own patterns of grieving as much features of our uniqueness as our fingerprints. Some people become quiet and contemplative; others throw themselves into a frenzy of work; yet others drink more, or cut themselves off from friends. Whatever the loss, they seem always to adopt the same response.

Some reactions are common to most of us. One is anger, bitterness, or resentment. We feel angry at what we regard as unnecessary delays in making the diagnosis or instituting treatment; anger that he smoked when he should have known what it would do to him; anger that he overworked and weakened himself; anger with God for not answering prayers; anger with those who were his friends and who now seem of no help; and so forth. The strength of this anger comes as a shock to many. They do not know how to handle it.

Caring and grieving are both exhausting. The tiredness which results comes as yet another surprise. For a while it is ignored but soon can no longer be, and it leads to frustration, frayed tempers, personal recrimination, and a deep sense of inadequacy.

One further feature both of grieving and the continuing exhaustion is the inability to cry and mourn as one would expect to. On some days there are no more tears; on others, a feeling almost of unwillingness to visit the hospital or to cope any longer at home. The carer may wonder whether this is an end to the love or a resentment against the dying one. It is quite normal—a feature of unutterable weariness. It can be so profound that when death comes at last, it seems almost an anti-climax and often there are no tears.

Immediately after the death, immensely painful as people expect it to feel, there is actually a sense of relief. The carer is surrounded by friends, is supported and understood. This sense of being loved continues for a few weeks until life begins to return to normal for everyone except the carer. Then the grief really cuts deep into the soul. As weeks become months, everyone else seems to behave as though the relative had never died, indeed as if he had never lived. The world goes on as before. His name is less often mentioned; fewer people express their sympathy and the carer finds himself or herself 'looking for', 'searching for' the one who has gone. Photographs are taken out and memories recalled. Hours are spent going through wardrobes of clothes, each potent stimulators of memories of happier days.

Unexpectedly grief takes on an additional feature. The dead one seems to be so near. He can be heard at night locking the door or putting the cat out; there is the smell of his tobacco or aftershave or her perfume as it wafts through the house. It sounds, as it always did, when he says he is going upstairs to bed . . . all in the mind, yes, but are these the dreaded hallucinations of a mental illness? Most emphatically not. They are quite normal, even though so frightening and unexpected. Many people experience them but because they are too frightened to speak of them for fear of being labelled 'mad', no-one does speak of them so no-one expects them. To make it worse, the one who is left sometimes begins to experience the same symptoms as the loved one who has died—the same palpitations or heartburn, the same breathlessness or changed bowel habit. Understandably this is alarming. Is this the beginning of the end? Most likely not but there is only one way to find out. Many experienced family doctors will have anticipated and made advance arrangements for thorough medical examinations of the carers but, if not, an appointment with the doctor should be made as soon as possible. Far better to have everything checked than to go on in stoical silence and isolation with secret fears breeding more fears.

These experiences—wondering if you're becoming mentally ill or wondering if you're developing the same illness as the one who has died—are understandably upsetting. Many other experiences can also upset and being forewarned of them may help. The apparent indifference of many friends and acquaintances always surprises and upsets. At first they cannot say or do enough. Within weeks they appear indifferent. It has to be remembered that many people simply do not know what to say and, rather than saying the wrong thing, end up by saying or doing nothing at all. Often the best intentioned words of comfort upset you. A friend might say of an old person who has just died 'Anyway he had a good innings', only to receive the heartbroken relative's reply 'Maybe, but I still want him here not dead!' 'It must have been very trying for you looking after him' may produce a surprisingly vehement denial when everyone knew that caring for him had indeed been frustrating and exhausting.

Another surprise is how easily the slightest thing can trigger a painful, tearful reaction, often months or even years after the death. Occasionally you can identify it—a familiar smell of perfume or cooking or tobacco, a much-loved piece of music, or a view from a beauty spot. Most times, no trigger can be identified. One minute you're busily occupied with something, the next torn apart with sadness and, as people always say, self-pity. The unpredictability makes people wonder if they will ever get over it.

In attempts to prevent such waves of grief and tears, some people not only stipulate 'no flowers or letters, please' but go to great lengths to clear the house of all removable mementos—clothes, papers, photograph albums, golf clubs, and all the other things we hoard in our own individual ways. It does not work! You cannot throw out memories, the way you can throw out a case of old clothes or a bag of clubs. People who try to do so often say how, later, they wished they still had a photograph or some little treasure. Of course there are a few grievers who not only throw nothing away, but keep a room exactly as it was with the open book, the papers untouched, the pen ready to use—a veritable museum or mausoleum. This is *not* a sign of healthy grieving.

It might be expected that for those with a faith this would be a comfort for them, somehow making sense of sorrow and pain. For many this is undoubtedly so and they must find it surprising that it is not always as helpful for others. The fact is that there are many who temporarily lose their faith. They feel cheated by God, their prayers that He would heal the loved one ignored. They resent it that a life of faith should be so rewarded and are often particularly upset by the insensitivity of many church friends. Eventually, sometimes years afterwards, faith gradually returns and God is forgiven.

If people can even blame God, how much more so can they blame others? The family doctor is often thanked profusely (like all the other caring professions) immediately after the death. Within a few months in the normal course of grief, he is seen in a new light and is criticized for being slow to diagnose the illness or refer to a specialist, reluctant to come out at night, or too hasty in offering a prescription. Interestingly, he is often held in slightly lower regard because of his newly revealed skill in telling lies! Once regarded as a pillar of honesty and integrity, he so readily agreed not to tell the truth to the patient. In fact, he told one 'white lie' after another at the request of the carer who now wonders if the doctor can ever be trusted again.

It is often asked how long 'normal' grief should last. It's tempting to reply rather facetiously 'How long is a piece of string?'—but there is some comfort in knowing that when you can still feel desolate more than a year after the death, your grief is normal. Perhaps more people should be advised that there are still many tearful, empty days 15 or 18 months later, after which it will very, very gradually, almost imperceptibly, improve. An illustration often used and found helpful is that of the operation scar. In the days and weeks immediately after an operation, it is exquisitely tender and all too obvious. As time passes it becomes less tender, less obvious, but never disappears no matter how many years pass. That is like grief, no more to be denied than that scar.

Appendix A: Useful equipment for home care

It should be noted that some of these items may only rarely be needed, others only by a few patients with very special needs. Many can be loaned, others may have to be purchased, but always the family doctor and nurse will advise.

Anti-pressure-sore mattresses (Spenco, Ripple, Ro-ho, and so on)
Audio-cassette/CD player with headphones or pillow phone
Backrest
Bedside fan
Commode (with or without a back, or arm-rests)
Delta (V-shaped) pillow
Deodorizer (ultraviolet light or charcoal, not aerosol)
Extra ice-trays for the refrigerator to ensure plentiful supply of ice-cubes and 'lollipops'
Liquidizer
Male urinal
'Monkey' pole
Nylon bed ladder
Remote-control TV set
Room-to-room intercom
Sheepskin mattress, real or synthetic
Special cutlery, plates, and mugs for the disabled
Vacuum flask (for hot or cold drinks)
Wine bottle cooler (not for wine but for other chilled drinks!)

Appendix B:
Useful organizations

ACT

(Action for the Care of Families whose children have life-threatening and terminal conditions)
Institute of Child Health, Royal Hospital for Sick Children, St Michael's Hill, Bristol BS2 8BJ. Tel. 0272 221556.

Parents and professionals may either telephone or write for information about support services and self-help groups, including children's hospices.

Association of Crossroads Care Attendant Schemes

10 Regent Place, Rugby, Warwickshire CV21 2PN.
Tel. 0788 573653.

This organization provides care attendants who come into the home to provide a break for the carers. Applications should be made either to the above address or to:

In Wales: *North Wales:* Crossroads, The North Wales Regional Office, 104 – 106 High Street, Mold, Clywd CH7 1VH.
Tel. 0352 750544.

South Wales: (Main Office) Crossroads Wales, 5 Coopers Yard, Trade Street, Cardiff CF1 5DF. Tel. 0222 222282.

In Scotland: 24 George Square, Glasgow G2 1EG.
Tel. 041 226 3793.

BACUP

3 Bath Place, Rivington Street, London EC2A 3JR.
Tel. 071 613 2121. Freephone outside London—0800 181199.
This organization helps patients and their families to cope with cancer. You may either telephone or write, and specially trained cancer nurses will provide information, support, and practical advice. In addition, the organization publishes many useful leaflets and booklets on different forms of cancer and their treatment.

Breast Care and Mastectomy Association of Great Britain

15/19 Britten Street, London SW3 3TZ. Helpline telephone number—071 867 1103.
In Glasgow: Suite 2/8, 65 Bath Street, Glasgow G2 2PS.
Tel. 041 353 1050.
In Edinburgh: 9 Castle Terrace, Edinburgh EH1 2DP.
Tel. 031 228 6715.
This organization, staffed partly by volunteers who have themselves had breast cancer, offers practical advice, information, and support to women concerned about breast cancer.

British Colostomy Association

15 Station Road, Reading, Berkshire RG1 1LG.
Tel. 0734 391537.
This is an information and advisory service, partly staffed by helpers who have themselves long experience of living with a colostomy, offering information, reassurance, and encouragement to patients with colostomies.

British Red Cross Society

9 Grosvenor Crescent, London SW1 7EJ. Tel. 071 235 5454.
(Local offices are listed in telephone directories and Yellow Pages.)
The British Red Cross offers a range of services of use to cancer patients, including the loan of wheelchairs and other pieces of equipment.

Cancerlink

17 Britannia Street, London WC1X 9JN. Tel. 071 833 2451.
This organization provides emotional support and information on all aspects of cancer to help patients, families, and friends and also acts as a coordinating resource for cancer support and self-help groups throughout the country. Enquirers in Scotland should contact Cancerlink Scotland, 9 Castle Terrace, Edinburgh EH1 2DP. Tel. 031 228 5557.

Cancer Relief Macmillan Fund

15/19 Britten Street, London SW3 3TZ. Tel. 071 352 7811.
Scottish office: 9 Castle Terrace, Edinburgh EH1 2DP.
Tel. 031 228 5557.
This major national charity helps patients and their families in many ways. It pump-primes the establishment of Macmillan Nursing Services, does much to encourage professional education in cancer care, and also has a patient grants department, providing financial help towards the cost of a wide range of needs for people with cancer. Applications for such grants are usually made by a home care nurse, a health visitor or community nurse, or a social worker.

Carers' National Association

In England: 22 Chilworth Mews, London W2 3RG.
Tel. 071 724 7776.
In Scotland: 11 Queen's Crescent, Glasgow G4 9AS.
Tel. 041 333 9495.
This organization provides information and support for people caring for patients at home and has a range of free leaflets.

Compassionate Friends

6 Denmark Street, Bristol BS1 5DQ. Tel. 0272 292778.
This is a befriending rather than a counselling service for parents whose child of any age has died from whatever cause. Information can be provided about local groups nationwide.

CRUSE Bereavement Care

CRUSE House, 126 Sheen Road, Richmond, Surrey TW9 1UR.
Tel. 081 940 4818.
In Scotland: 18 South Trinity Road, Edinburgh EH5 3PN.
Tel. 031 551 1511.
In Wales: Bryn Tirion, Churchill Close, Llanblethian, Cowbridge,
South Glamorgan CF7 7JH. Tel. 0446 775351.
This national organization for bereavement care and counselling
now has more than 200 branches throughout the United Kingdom,
their addresses and telephone numbers to be found in the directory
or Yellow Pages. Those in need may make direct approach to
a local branch or be referred or recommended by family doctor,
health visitor, Macmillan nurse, or friends. Most branches offer
one-to-one counselling either in the office or the person's home plus
social groups, children's counselling, and social activities.

Foundation for Black Bereaved Families

11 Kingston Square, Salter's Hill, London SE19 1JE.
Tel. 081 761 7228.
The foundation offers counselling, financial advice and support
for bereaved black people of Afro-Caribbean origin.

Gay Bereavement Project

Unitarium Rooms, Hope Lane, London NW11 8BS.
Tel. 081 455 8894.
This is a telephone helpline service for people bereaved by the
death of a partner of the same sex.

Hodgkin's Disease Association

PO Box 275, Haddenham, Aylesbury, Buckinghamshire HP17 8JJ.
Tel. 0844 291500.
As its name implies, this organization offers information and
emotional support for patients suffering from both Hodgkin's
disease and non-Hodgkin's lymphoma, as well as their families.

Hospice Information Service

St Christopher's Hospice, 51 – 59 Lawrie Park Road, Sydenham, London SE26 6DZ. Tel. 081 778 9252.

This service publishes an invaluable Directory of Hospice Services in the United Kingdom and Ireland, including basic details of hospices, home care teams, and hospital support teams.

Irish Cancer Society

5 Northumberland Road, Dublin 4, Ireland.
Tel. 010 3531 681855 (Freephone in Eire 1800 200300).

The freephone service offers information on all aspects of cancer from specially trained nurses. The society funds home care and rehabilitation programmes, offers support groups, and, on the request of the patient's doctor or public health nurse, a home night nursing service.

Leukaemia Care Society

PO Box 82, Exeter EH2 5DP. Tel. 0392 64848.

This society promotes the welfare of patients and their families suffering from leukaemia and allied blood disorders, offering support, friendship, information leaflets, and financial assistance.

Malcolm Sargent Cancer Fund for Children

14 Abingdon Road, London W8 6AF. Tel. 071 936 4548.

This organization is for young people under the age of 21 who have any form of cancer, whether they are in hospital or in their own home.

Marie Curie Cancer Care

28 Belgrave Square, London SW1X 8QG. Tel. 071 235 3325.

This is another major charity which not only runs several large hospices but also funds home nursing services and is active in professional education and cancer research.

National Association of Laryngectomee Clubs

Ground Floor, 6 Ricket Street, London SW6 1RU.
Tel. 071 381 9993.

This asssociation encourages rehabilitation, speech therapy, and social support and advises on special aids and equipment for laryngectomees.

National Holiday Fund for Sick and Disabled Children

Suite 1, Princess House, 1 – 2 Princess Parade, Dagenham, Essex RM10 9LS. Tel. 081 595 9624.

This fund provides holidays throughout the world for children between 8 and 18 with either chronic or terminal conditions.

Tak Tent

G Block, Western Infirmary, Dumbarton Road, Glasgow G11 6NT.
Tel. 041 332 2639.

This organization, which started in Glasgow, has now spawned groups throughout Scotland, offering emotional support, counselling, and information on cancers and treatments, as well as running various 'coping with cancer' courses.

Appendix C: Useful reading

Books for and about children

Burton, L. (1974). *Care of the child facing death*. Routledge and Kegan Paul, London.

Department of Social Work, St Christopher's Hospice (1991). *Your parent has died*. St Christopher's Hospice, 51 – 59 Lawrie Park Road, London SE26 6DZ.

Krements, J. (1983). *How it feels when a parent dies*. Victor Gollancz, London.

Mathias, B. and Spiers, D. (1992). *A handbook on death and bereavement. Helping children understand*. National Library for the Handicapped Child, Reach Resource Centre, Signpost Books, London. (A thoroughly comprehensive bibliography of all the major books and pamphlets to help grieving children.)

Snell, N. (1987). *Emma's cat dies*. Hamish Hamilton's Children's Books, 27 Wrights Lane, London W8 5TZ.

Williams, G. and Ross, J. (1983). *When people die*. Macdonald, Edinburgh.

Books about hospices and their work

Lewis, M. (1989). *Tears and smiles—the hospice handbook*. Michael O'Mara, London. (The definitive book for the layman on the hospice movement in Britain.)

Manning, M. (1984). *The hospice alternative—living with dying*. Souvenir Press, 43 Great Russell Street, London WC1B 3PA.

Stoddard, S. (1979). *The hospice movement—a better way of caring for the dying*. Jonathan Cape, London.

Zorza, R. and Zorza, V. (1980). *A way to die*. André Deutsch, London. (A moving account by two journalists of their daughter's death in a British hospice.)

Spiritual care

Ainsworth-Smith, I. and Speck, P. (1982). *Letting go—caring for the dying and bereaved.* New Library of Pastoral Care, SPCK, Holy Trinity Church, Marylebone Road, London NW1 4DU.

Cassidy, S. (1988). *Sharing the darkness: the spirituality of caring.* Darton, Longman, and Todd, London.

Casson, J. H. (1980). *Dying—the greatest adventure of my life.* Christian Medical Fellowship, 157 Waterloo Road, London SE1 8XN.

Chidwick, P. (1988). *Dying yet we live—our spiritual care of the dying.* Anglican Book Centre, Toronto, Canada.

Interdenominational Working Party (1991). *Mud and stars. The impact of hospice experience in the church's ministry of healing.* Sobell Publications, Sir Michael Sobell House, Churchill Hospital, Oxford OX3 7LJ.

Speck, P. (1978). *Loss and grief in medicine.* Baillière-Tindall, London. (A book written by a hospital chaplain, principally for professionals but richly rewarding for the lay reader.)

Twycross, R. G. (1982). *The dying patient.* Christian Medical Fellowship, 157 Waterloo Road, London SE1 8XN.

Books about grief

Collick, E. (1986). *Through grief—the bereavement journey.* Darton, Longman, and Todd, London.

Kubler-Ross, E. (1987). *On death and dying.* Tavistock, London. (This is one of the seminal works on the subject, written primarily for professionals but of value to all who care for the dying or are themselves grieving.)

Lewis, C. S. (1971). *A grief observed.* Faber and Faber, London. (A sensitive, personal account of his grief by C. S. Lewis, the well known author.)

Pincus, L. (1974). *Death and the family: the importance of mourning.* Faber and Faber, London. (This is a book about not only death and grief but also the dynamics of the family.)

Rando, T. A. (1988). *Grieving: how to go on living when someone you love dies.* Lexington (Macmillan), New York.

Richardson, J. (1979). *A death in the family.* Lion Publishing, Icknield Way, Tring, Herts. (This is a most valuable and comprehensive book, full of practical help and advice for carers.)

Sherr, L. (ed.) (1989). *Death, dying and bereavement.* Blackwell Scientific. (A book principally for the professionals but of value to all carers.)

Smith, K. (1979). *Help for the bereaved.* George Duckworth, 43 Gloucester Road, London NW1.

Staudacher, G. (1988). *Beyond grief: a guide for recovering from the death of a loved one.* Condor Books, Souvenir Press, 43 Great Russell Street, London WC1B 3PA.

Stedeford, A. (1994) (2nd Ed.). *Facing death: patients, families, and professionals.* Sobell Publications Oxford. (Written by a psychiatrist with long experience in this work, it provides the most comprehensive insight and understanding of what each person, whether patient or carer, experiences.)

Torrie, M. (1975). *Begin again: a book for women alone.* Aldine Paperback, J. M. Dent, London.

Whitacker, A. (ed.) (1984). *All in the end is harvest: an anthology for those who grieve.* Darton, Longman, and Todd, London.

Aiding better communications

Buckman, R. (1988). *I don't know what to say: how to help and support someone who is dying.* Papermac (Macmillan). (Written by an oncologist who has made a lifetime study of communications, this is an invaluable book for all caring for the dying.)

Stedeford, A. (1994) (2nd Ed.). *Facing death: patients, families, and professionals.* Sobell Publications Oxford. (Written by a psychiatrist with long experience in this work, it provides the most comprehensive insight and understanding of what each person, whether patient or carer, experiences.)

Sources of help

Cancer Relief Macmillan Fund. *Help is there: national contacts for people with cancer.* Cancer Relief Macmillan Fund, Anchor House, 15/19 Britten Street, London SW3 3TZ. (This leaflet gives information about 32 organizations in Britain, offering information, support, counselling, and grants for patients with the whole range of cancers.)

Quaker Social Responsibility and Education (1984). *A list of books and pamphlets to help Friends facing death or bereavement.* Quaker Social Responsibility and Education, Friend's House, Euston Road, London NW1 2BJ. (A list of useful organizations and books, many of which are included in this bibliography.)

Speechly, V. and Rosenfield, M. (1992). *Cancer information at your fingertips: the comprehensive care reference book for the 1990s.* Class, London. (This invaluable book lists every major organization and source of information likely to be needed by the cancer patient and those involved in his care.)

Index